Weathering the Storm

Smallholder farmers face relentless challenges from pests, diseases, and the severe impacts of climate change across developing countries. Each year, they endure droughts and floods, thrusting them into a vicious cycle of poverty and distress. *Weathering the Storm: Farmer Resilience and Adaptive Strategies to Crop Losses* provides a vital exploration of these issues, offering insights into the struggles and resilience of farmers in the wake of devastating crop losses.

Drawing on data from 1,440 farmers who faced severe crop losses in three states namely Telangana, Andhra Pradesh, and Karnataka of India, the book uncovers:

- **Staggering Losses:** Crop losses range from 30% to 100%, causing financial damages of three to five lakhs rupees per farmer.
- **Government Response:** A critical analysis of various assistance programmes, highlighting the effectiveness of schemes like the Pradhan Mantri Fasal Bima Yojana (PMFBY) and YSR free crop insurance.
- **Resilience Strategies:** Insights into the self-adopted strategies farmers use to cope with challenges, alongside governmental support.
- **Effective Governance:** Examination of Andhra Pradesh's decentralized governance model as a best practice for rapid response.
- **Farmers Distress Index (FDI):** Introduction of an innovative forewarning system to identify farmers and areas in urgent need of intervention.

A. Amarender Reddy, an agricultural economist and Joint Director of Policy Support Research at the ICAR-National Institute of Biotic Stress Management (ICAR-NIBSM) in Raipur, explores the intricate challenges of agrarian distress and champions holistic solutions. This insightful book is a must-read for anyone passionate about the future of agriculture and the resilience of rural livelihoods.

Weathering the Storm

Farmer Resilience and Adaptive Strategies to Crop Losses

A. Amarender Reddy
Joint Director (School of Crop Health Policy Support Research)
ICAR-National Institute of Biotic Stress Management
(ICAR-NIBSM), Raipur

CRC Press
Taylor & Francis Group
Boca Raton London New York

CRC Press is an imprint of the
Taylor & Francis Group, an **informa** business

Designed cover image: Author's own image

First edition published 2025
by CRC Press
2385 NW Executive Center Drive, Suite 320, Boca Raton FL 33431

and by CRC Press
4 Park Square, Milton Park, Abingdon, Oxon, OX14 4RN

CRC Press is an imprint of Taylor & Francis Group, LLC

ISBN: 9781032947488 (hbk)
ISBN: 9781032950204 (pbk)
ISBN: 9781003582878 (ebk)

DOI: 10.1201/9781003582878

Typeset in Times
by codeMantra

Contents

प्रो. रमेश चन्द
सदस्य
Prof. Ramesh Chand
MEMBER

सत्यमेव जयते

भारत सरकार
नीति आयोग, संसद मार्ग
नई दिल्ली-110 001
Government of India
NATIONAL INSTITUTION FOR TRANSFORMING INDIA
NITI Aayog, Parliament Street
New Delhi-110 001
Tele. : 23096756, 23096774 Fax : 23730678
E-mail : rc.niti@gov.in

Foreword

It is with great pleasure that I write the foreword for the book, *Weathering the Storm: Farmer Resilience and Adaptive Strategies to Crop Losses*, authored by Dr. A. Amarender Reddy. The book provides an insightful and timely analysis of the challenges faced by farmers due to vulnerability of crop output to vagaries of nature in Telangana, Andhra Pradesh, and Karnataka. The manuscript documents the empirical evidence around the thrips attack of 2021–22 which devastated chilli crop in the region, and narrates the plight of farmers who faced this. The book provides a comprehensive assessment of the economic and socio-psychological impacts of crop losses, along with the adaptive strategies followed by farmers and the role of government support in mitigating the risks.

Rising frequency of pest outbreaks, extreme weather events, and other uncertainties pose serious threats to rural livelihoods. This book goes beyond mere documentation of crop losses—it offers a nuanced exploration of farmers' coping mechanisms, including their financial, technological, and socio-psychological adaptations. It also evaluates the effectiveness of interventions by local, state and central governments, providing valuable insights into the support systems that can help farmers build resilience.

Micro-level studies like this are critical for evidence-based policymaking. By highlighting the ground realities of farming communities, the authors offer policymakers a data-driven foundation to design agricultural policies targeted at real world problems and their solution. The book's findings will be of immense value to policymakers, researchers, and development practitioners seeking to enhance agricultural resilience and safeguard the livelihoods of farmers. It is also pertinent to highlight that strict monitoring of quality of farm inputs is critical as the reports of supply of substandard inputs especially agro-chemicals are appearing frequently. Excessive reliance on agro-chemicals to control insect and pest exacerbate the situation and calls for alternative methods like IPM and crop diversity to keep check on pests' infestation. High value crops like chilli promises high return associated with high production and market risk. A well-diversified farming system is also found effective tool to address high risk due to high concentration under one or two crops.

I congratulate the author for in depth analysis and rigour used to address chronic and widespread problem of risk in crop production. I also commend Taylor & Francis for publishing this significant contribution. I am confident that this book will serve as a valuable resource for agricultural economics students and faculty and also policy makers who are interested in Indian agriculture.

[Ramesh Chand]

Place: New Delhi
Date : March 24, 2025

स्वच्छ भारत
एक कदम स्वच्छता की ओर

Preface

Farmers in India face a myriad of challenges stemming from natural calamities such as droughts, floods, and pest outbreaks, compounded by the effects of climate change and neoliberal agricultural policies. These policies, marked by privatization and free market forces, expose farmers to production and price risks, exacerbating distress in recent years. Farmers' distress encompasses various hardships across financial, social, and emotional domains, with small and marginal landholders and tenant farmers bearing the brunt of vulnerability.

The hazards of droughts, floods, severe pest attacks, and price fluctuations significantly impact crop yields, profitability, and livelihoods, particularly for small-scale farmers. Root causes of distress include inadequate rainfall, lack of irrigation facilities, and outbreaks of pests and diseases, leading to crop failure and subsequent financial losses. Moreover, over reliance on purchased inputs like pesticides and fertilizers, alongside soil degradation and unsustainable agricultural practices, further diminishes profitability for rainfed and smallholder farmers.

These challenges perpetuate a cycle of poverty and indebtedness, leaving farmers grappling to meet basic needs. Addressing agrarian distress necessitates a holistic approach that tackles underlying causes. While government schemes like crop insurance and direct benefit transfers offer support, their effectiveness is hindered by delays and inadequate responsiveness to external shocks such as significant crop losses. Urgent reforms are imperative to ensure timely and effective assistance to farmers in distress.

The cultivation of chilli crops represents a significant commercial venture in the southern states of India, characterized by its high-input, high-output nature, coupled with inherent high-risk factors. Typically, farmers invest approximately Rs. 1.25 lakhs per acre in chilli cultivation, with the potential to reap profits of about Rs. 3 lakhs per acre under favourable conditions. However, the failure of the crop results in a complete loss of investment. This book critically examines the challenges faced by farmers in the aftermath of severe losses caused by a thrips outbreak in 2021–2022 across three states: Telangana, Andhra Pradesh, and Karnataka.

The outbreak of thrips, specifically *Thrips parvispinus* (Kudo), in chilli fields has sparked discussions on farmers' distress and the efficacy of various state government programmes aimed at alleviating this distress. *T. parvispinus* is a cosmopolitan species of quarantine significance, with its geographic distribution expanding drastically over the past two decades, affecting crops from Thailand to Australia and beyond. In India, this pest was initially reported on papaya in Bengaluru, Karnataka, in 2015, but, since 2021, it has emerged as a significant threat to chilli crops, wreaking havoc in fields.

The reported crop losses due to thrips infestation ranged from 70% to 100%, resulting in monetary losses exceeding 3–5 lakhs per farmer. The book critically analyses and illustrates the socio-economic status of farmers before, during, and after this widespread outbreak, examining the impact on their incomes and the responses of different state governments to this agrarian crisis. Given India's federal structure, with agriculture being a state subject, different states implement varying programmes and schemes to aid farmers. The book also investigates the effectiveness of these government initiatives.

Field surveys conducted in 2022 gathered data for the crop years 2020–2021 and 2021–2022. The former was characterized as a normal year with average profits across major crops in all three states, while the latter witnessed a bad year with an average crop loss of 30–35%, particularly affecting chilli crops. To comprehend the diverse adaptation and coping strategies employed by farmers, the book delves into their socio-economic conditions, crop conditions, losses, and benefits received from different schemes during both normal and adverse years, drawing insights from a survey of 1,440 farmers.

The book's chapterization unfolds with a comprehensive exploration of various facets of agricultural distress and resilience. The opening chapter, "Farmers' Journey through Uncertainty," immerses readers into the challenges faced by farmers amidst agrarian distress, elucidating the fundamental characteristics of Indian agriculture. It meticulously examines the status of farmer distress and underscores the pivotal role of governmental interventions in alleviating these hardships.

Following this, "Exploring the Unpredictable Agriculture" delves into the intricacies of risk and variability in agricultural yields, prices, and returns at the district level. The findings underscore the wide-ranging disparities across districts, prompting a call for in-depth analyses of the multifaceted risk factors exacerbating farmer distress.

The third chapter, "Strategies for Gathering Evidence," delineates the meticulous methodology employed in the study. It articulates the rationale behind selecting specific states, outlines the extent of crop loss, and underscores disparities in policy frameworks across regions. Additionally, it delves into the meticulous process of selecting and interviewing sample farmers to gather comprehensive data on crop losses, farmer distress, and governmental responses.

The subsequent chapter, "Farmer Assets and Wealth Dynamics," presents insights gleaned from field observations and farmer interviews. It sheds light on the distribution of assets and wealth across diverse farming groups, unravelling variations in adoptive capacities among these groups.

Next, "Story of Farmer's Land" delves into the intricate dynamics of land ownership and tenancy across different social strata and regions. It sheds light on the struggles faced by small farmers in the face of land shortages, often leading them to resort to land rental – a practice that becomes particularly burdensome during years of crop loss.

The sixth chapter, "Exploring Farmers Income: A Tale of Harvesting Hope," provides a nuanced examination of how farmers' incomes fluctuate between normal and abnormal years, such as 2021 and 2022. It delves into the diverse sources of income for farmers and illuminates how they adapt to crop losses by augmenting their incomes through various means such as MGNREGA works, off-farm, non-farm casual employment, and migration. Moreover, the chapter assesses the extent to which poverty is exacerbated by crop losses and identifies farmers who are particularly vulnerable and struggle to stabilize their incomes.

The seventh chapter, "Patterns in Household Savings," elucidates the saving patterns of households across all farm size groups during both normal and abnormal years, particularly focusing on heavy crop loss years. It delves into the challenges farmers face in saving money amidst income uncertainty and shares stories of farmers who have devised innovative ways to save despite adversities.

Chapter 8, "Debt Burden of the Households," delves into the profound struggles farmers encounter in managing debt across different farming communities in India. It paints a vivid picture of how loans, both formal and informal, serve as lifelines for farmers, yet simultaneously pose significant hardships, especially during crop loss disasters.

The ninth chapter, "Investigating into Crop Losses and Recovery," explores the root causes of farmers' substantial debt burdens, especially following crop failures attributed to unpredictable weather, pests, soil problems, and inadequate storage facilities. It uncovers how repeated crop failures compel farmers to borrow extensively from informal sources to endure. While governmental efforts such as crop insurance and improved farming methods aim to alleviate these burdens, their effectiveness remains limited, except in cases like Andhra Pradesh, where initiatives like village secretariats and Rythu Bharosa Kendras (RBKs) have established transparent and responsive local governance systems.

Chapter 10, "Barriers to Adopting Antifragile Farming Practices," sheds light on the obstacles farmers encounter in embracing resilient farming methods post-crop loss. Despite these challenges, farmers persist in seeking solutions, exploring sustainable techniques like water-saving crops, conservation agriculture, and improved irrigation. However, the chapter highlights how small farmers face difficulties in adopting these methods due to high costs, debt burdens, insufficient capital, and societal pressures, despite commendable governmental initiatives.

Chapter 11, "Understanding Farmers' Perception of Risk," delves into the intricate ways in which farmers perceive and navigate risks associated with adopting new farming methods. It underscores the profound influence of past experiences, social norms, economic factors, and religious beliefs on farmers' risk perceptions. The chapter highlights farmers' cautious approach towards embracing new technologies, driven by fears of failure and societal judgment. It emphasizes the significance of comprehending farmers' risk perceptions to devise targeted interventions and support systems that cater to their specific needs and challenges in adopting innovative farming practices.

The 12th chapter, "Psychological Stress: Another Burden of Farmers," sheds light on the often-overlooked psychological toll experienced by farmers, exacerbated by indebtedness and crop losses. It elucidates how financial stress renders farmers incapable of adequately caring for their families, compounded by inadequate support from nuclear families, further intensifying mental strain. The chapter delves into various factors such as past experiences, financial strain, societal expectations, and discrimination, illustrating their diverse impacts on farmers' mental well-being. Despite these immense pressures, most farmers persevere in their endeavours to support their families and uphold their way of life.

Chapter 13, "Crop Yield Losses and Economic Costs," portrays farming as a challenging business endeavour, highlighting the complexities involved in achieving profitability amidst crop losses and escalating expenses. It emphasizes the relentless pressure farmers face as expenses surge while product prices stagnate. The chapter underscores the significance of insurance in mitigating financial risks associated with crop losses. Furthermore, it elucidates how farmers across different regions employing diverse farming methods contend with varying levels of crop losses, thereby impacting their earnings. Additionally, the chapter addresses challenges related to excessive pesticide and fertilizer usage, as well as issues pertaining to seed storage and produce transportation. Ultimately, it advocates for increased support and understanding of the farming sector, acknowledging its pivotal role in sustaining both society and the economy.

Chapter 14, "Bridging the Gaps: Government Welfare Schemes and Their Reach," scrutinizes the challenges farmers encounter in accessing government assistance, particularly during normal and heavy crop loss years. It underscores how certain groups receive more support than others due to factors like complicated paperwork and corruption, leaving many poor and tenant farmers struggling to benefit. The chapter emphasizes the necessity of targeted programmes for marginalized groups and advocates for better outreach, drawing inspiration from successful models like those in Andhra Pradesh. It stresses the urgent need to bridge the gap between government initiatives and the farmers most in need, emphasizing the pivotal role of effective local governance in maximizing government support for farmers' welfare.

Chapter 15, "Effects of Farmer Loan Waivers," explores the unequal outcomes of farmer loan waivers in India, highlighting their shortcomings and complications for different farmer groups. Despite their intent to aid struggling farmers, loan waivers are found to be exploited for political gains, exacerbating biases and discriminations among farmers. Marginalized social groups receive minimal waivers, and small farmers, primarily tenant farmers, often miss out due to their reliance on money lenders. The chapter concludes by advocating for fairer policies to assist all types of farmers equitably.

Chapter 16 introduces the Farmers Distress Index (FDI), a simplified tool designed to forewarn various stakeholders, including the government, about the severity of farmers' distress based on seven pillars. This index, developed from 21 standardized variables, enables swift intervention by identifying areas with high distress scores. It proposes a scalable framework for FDI implementation at sub-district levels to facilitate timely government support where most needed.

Lastly, Chapter 17, "Scaling up Antifragile Strategies," offers concluding insights into India's agricultural distress, proposing strategies to address root causes and bolster resilience in farming communities. It emphasizes scaling up monitoring frameworks like the FDI to anticipate and mitigate the increasing frequency of crop losses and other calamities. By tracking standardized variables, governments can swiftly intervene to alleviate farmers' distress, ensuring the overall well-being of rural communities.

The study's major findings underscored the inadequacy of existing government schemes in responding effectively to frequent crop losses induced by natural calamities. Farmers' incomes plummeted during adverse years, exacerbating indebtedness and distress. Despite some effective schemes like crop insurance, challenges in implementation and coherence persisted across different government initiatives. Lessons from successful models in Andhra Pradesh underscored the importance of decentralized governance structures and streamlined procedures in enhancing government support for farmers. By adopting tools like the FDI and implementing medium-term interventions, governments can address farmers' distress more proactively, safeguarding rural livelihoods and fostering agricultural resilience.

A. Amarender Reddy
Joint Director, Policy Support Research
ICAR-National Institute of Biotic Stress Management (ICAR-NIBSM), Raipur

About the Author

A. Amarender Reddy holds a Ph.D. in Agricultural Economics from the ICAR-Indian Agricultural Research Institute (ICAR-IARI), New Delhi, and has over 24 years of experience in agricultural policy research, monitoring, and evaluation. He has worked with ICAR, ICRISAT, ASCI, MANAGE, and BIRD, contributing to the assessment of agricultural and rural development schemes.

As Director (Monitoring and Evaluation) at MANAGE, Hyderabad, he led multidisciplinary teams in evaluating flagship government programmes under the Ministry of Agriculture & Farmers Welfare.

Dr. Reddy has published over 100 research articles in reputed journals such as *Economic and Political Weekly*, *Food Policy*, and *Food Security*. He is also a reviewer and editor for international journals such as *World Development*, *Water Policy*, and *Journal of Decision Sciences*.

A Fellow of the Indian Society of Agricultural Economics, Mumbai, Dr. Reddy has received fellow and recognition awards from the Indian Society of Pulses Research & Development (ISPRD), Kanpur. He led research and evaluation studies funded by JBIC, NABARD, NIRD&PR, and the erstwhile Planning Commission on agrarian distress and farmer resilience.

His broad vision and practical experience in policy formulation and impact assessment add depth and credibility to this book, offering valuable insights into strategies and adaptation strategies to crop losses and building farmer resilience.

1 Introduction
Farmers' Journey through Uncertainty

Agriculture is the most important sector in the Indian economy providing livelihood support to about two-thirds of the Indian population despite its declining contribution to the overall economy. It is the world's largest producer of milk, jute, and pulses and the second-largest producer of rice, wheat, vegetables, sugarcane, cotton, fruit, and groundnut. It is also among the leading producers of spices, fish, poultry, livestock, and plantation crops (FAO, 2024). Further, the diversification of agricultural livelihoods through the allied sectors such as animal husbandry, forestry, and fisheries has enhanced livelihood opportunities and strengthened resilience too. Thus, agriculture and its allied sectors serve as the primary means of sustenance in India, particularly in rural areas. But, despite achieving food sufficiency in production and even exporting food, a new problem has emerged – "agrarian distress" or "farm distress" (NITI Ayog, 2020). That means farmers are facing difficulties. Desertification, land degradation, and water scarcity are the major threats to the country's agriculture.

To emphasize the severity of agrarian distress in India, one must consider the compounded effects of climate change, environmental degradation, and long-standing neglect of agricultural investments. Being the most climate-dependent economic activity, it is highly vulnerable to risks and uncertainties that threaten livelihoods (Stojanov et al., 2016). Research shows that increasingly erratic weather patterns, such as frequent droughts, floods, rising temperatures, soil erosion, and deforestation, are exacerbating the challenges faced by Indian agriculture (Reddy & Mishra, 2010; Singh & Singh, 2017). These climatic events often result in crop failures, due to either excessive rainfall or drought, and, even when weather conditions are favourable, low agricultural productivity often leads to depressed market prices (Deshpande, 2002). The resultant economic strain has left small and marginal farmers, who form the majority of India's agricultural workforce, especially vulnerable to these risks (NABARD, 2018).

The neglect of both public and private investment in agriculture over the decades has deepened this crisis, contributing to severe agrarian distress that continues to threaten the livelihoods of farming communities across India (Dev, 2008). The impact of these issues is evident in the widespread agrarian crises of recent decades, which persist with varying intensity across different regions (Swaminathan, 2010). Changing weather patterns disproportionately affect poor and marginalized farmers, leading to reduced food production and increased food prices, which many farmers can no longer afford (Patnaik, 2004). This, in turn, exacerbates poverty and limits the options for maintaining environmentally sustainable livelihoods, a vicious cycle

DOI: 10.1201/9781003582878-1

underscored by the tragic rise in farmer suicides across the country (Nagraj et al., 2014).

This crisis not only highlights the vulnerability of Indian farmers but also underscores the urgent need for comprehensive policy interventions to address the structural and environmental challenges facing the agricultural sector.

ROOTS OF AGRARIAN DISTRESS AND ITS SYMPTOMS – LET'S DIVE INTO THE WHY AND WHAT

Agrarian distress, a term that emerged in India in the early 1990s (Kandlur et al., 2022), is like a puzzle with many pieces. Farmers were facing challenges, like low incomes and debts, because they lacked resources, dealt with unpredictable markets, and faced high costs (Southard and Randell, 2022). Marginal and small farmers and farmers operating on rented land often feel more insecure (Jha et al., 2018). This distress connects to old issues in how land is used, and changes in technology have made things tricky, bringing in worries about food security. As time went on, trade opened, and the world's markets affected our farmers. Despite more money being put into farming, the safety nets were not strong enough, leading to agricultural disasters. With more people needing food, there is now a gap between what is needed and what is available. This mess is making it harder for farmers, especially as the number of workers in rural areas grows. Agrarian distress shows up in two ways – it affects the whole farming sector and hits individual farmers and their families. Signs of this problem include farmers getting into debt, diverting to other work, not having enough to eat, selling off assets to meet household expenditures, trying risky things, crops failing, and, sadly, more farmers attempting suicide (NITI Ayog, 2020). In a nutshell, agrarian distress is like a tough journey for farmers, with challenges at every turn, affecting not just their fields but their lives.

FARMER'S VOYAGE UNDER UNCERTAINTY

So, the picture is this – in recent years, everyone has been talking about a big problem called "agrarian distress" in India (Bhoi and Dadhich, 2019; Vasavi, 1999, 2009; Rao and Suri, 2006; Deshpande, 2002; Vyas, 2004; Posani, 2009; Chand, 2012). Farmers are going through a really tough time, and it's hitting them hard. Agriculture used to be the main income source for many, especially smaller farmers, but not anymore (Bhoi and Dadhich, 2019). More and more people are leaving farming to find other ways to make a living. Even though fewer households in rural areas depend on farming compared to before, it is still very important for many families (NSO, 2020). But lots of issues are making farming tricky. Farmers are drowning in debt, crops are not doing well, and as a result they are struggling with high expenses. On top of that, the weather is acting up with things like droughts and weird temperature changes (Ray et al., 2019; Rohini et al., 2016; Sharma and Majumder, 2017), making it even tougher for farmers who rely on rain for their crops (Rao et al., 2016). So, the farmers not making enough money have made things tougher and put them in a tight spot with serious money concerns (NSSO, 2014). Now, there's a financial organization called NABARD that surveyed in 2018 and found that almost half of the rural households are into agriculture, and about 35% of their income comes from it. But the sad part is – the income gap between farmers and non-farmers is growing, causing various social problems. It's gotten so tough for some farmers that they're resorting to something heartbreaking, that is suicide. The number of farmer suicides in India has become the highest in the world (Basha, 2018). In the old days, farmers had smart ways to deal with problems like sharing resources, swapping food, and using smart techniques to save soil and water. But now, with modern technologies and big changes in agriculture, those strategies are fading away. That's where the government comes in – it needs to create big plans to help farmers deal with these challenges and find new ways to make farming work better. It's like trying to fix a puzzle so that farmers can have a better and more stable life.

Governments Role: How Do Governments Help the Farmers?

If we imagine being a farmer who suddenly faces a disaster like a flood or a drought or cyclone that wipes out all the crops, it will be very scary. So, here's where the government steps in with something called "crop insurance," which is like a magical shield for farmers (Bahinipati et al., 2021; Cole and Xiong, 2017). This insurance helps them when their crops get damaged, making sure they still get some money even during tough times. It's not just about money – crop insurance lets farmers keep on farming even after a bad year. With crop insurance, farmers feel more secure just by paying a small amount of money called a premium. This can lead to better profits in the long run. Now, in India, the government has been working on different crop insurance plans to support farmers and protect them from crop losses. But not everyone agrees on how well these plans work, and there is a bit of debate among researchers. Some studies say they're great, bringing positive benefits (Cariappa et al., 2020), while others argue that in certain areas, they might not be effective (Cole and Xiong, 2017; Reddy et al., 2020). Besides crop insurance, the Indian government has other tools to help farmers, like compensating for crop loss, forgiving debts, rescheduling loans, and making sure farmers get a fair price for their crops (called minimum support price or MSP). But, among all these tools, crop insurance takes the spotlight. So, in a nutshell, the government is trying different tricks to make it even stronger and make sure the farmers can come back after tough times and keep our food supply steady.

Evolution of Crop Insurance in India

Crop insurance in India is a journey that started in the 1900s as a farmer's protective shield against crop disasters. Over the century, it has been changing, growing, and becoming even better at helping farmers.

Early on, in the pre-independence era a rain insurance idea turned up in Mysore State in 1915, but it did not take off. After India gained independence, crop insurance discussions heated up, and, by 1972, the first programme was launched in Gujarat. However, in 1979, the Pilot Crop Insurance Scheme took the spotlight, covering various crops. Then, in 1985, the Comprehensive Crop Insurance Scheme came into force, followed by the more improved National Agriculture Insurance Scheme in 1999. The journey continued with experiments, and, in 2010, the Modified National

Agricultural Insurance Scheme arrived, bringing in many changes like actuarial premiums and relief for farmers. In 2016, there was a fresh start with the Pradhan Mantri Fasal Bima Yojana (PMFBY), making sure farmers have the best protection for their crops.

PMFBY – THE FARMERS' SAFETY PLAN

It's like a safety net, giving farmers protection against crop troubles at a very affordable cost. Farmers pay a small percentage, and the government shares the rest. It's like an all-in-one insurance for their crops covering weather issues, pests, and post-harvest losses. But it doesn't cover everything – no war, nuclear stuff, or theft. The cost of premium varies for different crops and different seasons – Kharif (monsoon) season, which is 2% of the crop value; for others during the Rabi (winter) season, it is 1.5%; and for commercial crops, it is 5%. The government helps with a subsidy. Despite all these efforts, farmers still face challenges.

REFERENCES

Bahinipati CS, Kumar V, Viswanathan PK. (2021). An Evidence-Based Systematic Review on Farmers' Adaptation Strategies in India. *Food Security*, 13(2), pp. 399–418.

Basha PC. (2018). Farmers Suicide in India - Causes and Remedies. *International Journal of Research in Economics and Social Sciences*, 8(1), pp. 428–438.

Bhoi BK, Dadhich CL. (2019). *Agrarian Distress in India: Possible Solutions*. Indira Gandhi Institute of Development Research, Mumbai. Working Paper Series (2019-017).

Cariappa AA, Mahida DP, Lal P, Chandel BS. (2020). Correlates and Impact of Crop Insurance in India: Evidence from a Nationally Representative Survey. *Agricultural Finance Review*, 81(2), pp. 204–221.

Chand R. (2012). Development Policies and Agricultural Markets. *Economic and Political Weekly*, 47(52), pp. 53–63.

Cole SA, Xiong W. (2017). Agricultural Insurance and Economic Development. *Annual Review of Economics*, 9, pp. 235–262.

Deshpande RS. (2002). Suicide by Farmers in Karnataka: Agrarian Distress and Possible Alleviatory Steps. *Economic and Political Weekly*, 37(26), pp. 2601–2610.

Dev SM. (2008). Agriculture: Absence of a Big Push. *Economic and Political Weekly*, 43(26–27), pp. 33–36.

Food and Agriculture Organisation. (2024). FAO in India: India at a glance. https://www.fao.org/india/fao-in-india/india-at-a-glance/en/

Jha CK, Gupta V, Chattopadhyay U, Amarayil Sreeraman B. (2018). Migration as Adaptation Strategy to Cope with Climate Change: A Study of Farmers' Migration in Rural India. *International Journal of Climate Change Strategies and Management*, 10(1), pp. 121–141.

Kandlur R, Sardana S, Richardson-Vejlgaard R. (2022). The Agrarian Distress: Factors Explaining the Will to Live among Rural and Distressed Family Farmers. *Psychiatry Research Communications*, 2(1), p. 100019.

NABARD. (2018). *NABARD All India Rural Financial Inclusion Survey 2016–17*. NABARD. https://www.nabard.org/auth/writereaddata/tender/1608180417NABARD-Repo-16_Web_P.pdf

Nagaraj K, Sainath P, Rukmani R, Gopinath R. (2014). Farmers Suicides in India: Magnitudes, Trends, and Spatial Patterns, 1997–2012. *Review of Agrarian Studies*, 4(2), available at http://ras.org.in/farmers_suicides_in_india

National Sample Survey Organisation (NSSO). (2014). *Key Indicators of the Situation of Agricultural Households in India*. Ministry of Statistics & Programme Implementation, Government of India.

National Statistical Office. (2020). *Periodic Labour Survey (2018–19) Annual Report*. Ministry of Statistics and Programme Implementation, https://www.mospi.gov.in/sites/default/files/publication_reports/Annual_Report_PLFS_2018_19_HL.pdf

NITI Ayog. (2020). *NITI Ayog, Annual Report 2019–20*. NITI Ayog. https://www.niti.gov.in/sites/default/files/2020-02/Annual_Report_2019-20.pdf

Patnaik U. (2004). *The Republic of Hunger*. Public lecture, New Delhi.

Posani B. (2009). *Farmer Suicides and the Political Economy of Agrarian Distress in India*. Development Studies Institute Working Paper Series (09-95).

Rao CR, Raju BMK, Rao AS, Rao KV, Rao VUM, Ramachandran K, Venkateswarlu B, Sikka AK, Rao MS, Maheswari M, Rao CS. (2016). A District Level Assessment of Vulnerability of Indian Agriculture to Climate Change. *Current Science*, 10, pp. 1939–1946.

Rao PN, Suri KC. (2006). Dimensions of Agrarian Distress in Andhra Pradesh. *Economic and Political Weekly*, 41(16), pp. 546–1552.

Ray LK, Goel NK, Arora M. (2019). Trend Analysis and Change Point Detection of Temperature over Parts of India. *Theoretical and Applied Climatology*, 138(1–2), pp. 153–167.

Reddy DN, Mishra S. (2010). *Agrarian Crisis in India*. Oxford University Press.

Reddy AA, Raju SS, Bose A. (2020). Farmers' Income, Indebtedness and Agrarian Distress in India. *The Microfinance Review*, 12(1), pp. 20–38.

Rohini P, Rajeevan M, Srivastava AK. (2016). On the Variability and Increasing Trends of Heat Waves over India. *Scientific Reports*, 6, p. 26153.

Sharma S, Mujumdar P. (2017). Increasing Frequency and Spatial Extent of Concurrent Meteorological Droughts and Heatwaves in India. *Scientific Reports*, 7(1), pp. 1–9.

Singh, R., & Singh, G. S. (2017). Traditional Agriculture: A Climate-Smart Approach for Sustainable Food Production. *Energy, Ecology and Environment*, 2, pp. 296–316.

Southard EML, Randell H. (2022). Climate Change, Agrarian Distress, and the Feminization of Agriculture in South Asia. *Rural Sociology*, 87(3), pp. 873–900.

Stojanov R, Kelman I, Ullah AKM, Duží B, Procházka D, KK. (2016). Local Expert Perceptions of Migration as a Climate Change Adaptation in Bangladesh Blahůtová. *Sustainability*, 8(12), p. 1223.

Swaminathan MS. (2010). *From Green to Evergreen Revolution: Indian Agriculture Performance and Challenges*. Proceedings of the Indian National Science Academy.

Vasavi AR. (1999). Agrarian Distress in Bidar: Market, State and Suicides. *Economic & Political Weekly*, 34(32), pp. 2263–2268.

Vasavi AR. (2009). Suicides and the Making of India's Agrarian Distress. *South African Review of Sociology*, 40(1), pp. 94–108.

Vyas VS. (2004). Agrarian Distress: Strategies to Protect Vulnerable Sections. *Economic and Political Weekly*, 39(52), pp. 5576–5582.

2 Exploring the Unpredictable Agriculture

On an August afternoon in Paluvai village, Andhra Pradesh, four farmers were sitting outside a church, worried about the ongoing drought. They were speaking about life without hope and thinking of leaving the village for a city. Veeraswamy, a small farmer, planted cotton in May after some unexpected rain, but his crops failed due to the heat that followed. Their village, known for chilli and cotton crops, relies heavily on rain, which has been scarce that year. The delay in rainfall has left reservoirs at half their usual levels. Climate change made the drought worse. Veeraswamy, burdened by debt, sees no hope of recovering his losses. Farmers like Dasari Bujji also facing similar struggles, and many are considering moving to cities for work. In the past three years, ten debt-ridden farmers have committed suicide in their village and many more have left for cities (Sravani, 2023). This recent story in Andhra Pradesh highlights the unpredictability of agriculture, making us think deeply about the risks farmers face. Recognizing that long-term sustainable solutions require a multidisciplinary approach (Khatri et al., 2024), we, as researchers committed to understanding the improved agricultural landscape of our country, formed a group and indulged ourselves in exploring the unpredictable world of agriculture. We moved to study the instability of crop yields in three states of India, Andhra Pradesh, Karnataka, and Telangana, over the past six decades. It was a tough task, but we were determined to understand the unpredictable world of agriculture and its impact on the lives of farmers in rural areas.

So, we went deeper to examine the rollercoaster of instability in crops. For that, we took a journey back in time through the history of farming – the Green Revolution, when farming was changing fast; the Liberalization Era, when policies were being relaxed; and the Post-Liberalization Period, when things settled down a bit. We collected a data repository from a reputed organization named the International Crops Research Institute for the Semi-Arid Tropics on how much land was used (area), how many crops were grown (production), how well they grew (yield), and even how much money farmers made from selling them (farm harvest price or FHP). We peered into the highs and lows of multiple crop production, analysing its impact on the farmer's income.

After months of gathering data, we discovered that farming is like a rollercoaster ride for farmers where crop production and profits fluctuate depending on changes in prices, weather, water availability, and how products are sold (Reddy, 2006). And the fluctuations make the journey quite bumpy. These fluctuations made the situation tougher for farmers to decide where to invest their money and simultaneously hindered the agricultural sector's potential to contribute to economic growth. To deal with farming uncertainties and come up with good plans, it was important to understand how

DOI: 10.1201/9781003582878-2

unstable things are at different stages and how they have changed over the years (Reddy, 2012). While some past studies on agriculture suggested looking at how crops grow, how unstable their growth is, and how they respond to changes as technology evolves (Senapatia and Goyari, 2019), many have looked only at the overall, without paying much attention to how they affect the farmers' wallets (Hazell, 1982; Dev, 1987; Sharma et al., 2006).

HISTORICAL DIFFERENCES IN ANDHRA PRADESH, TELANGANA, AND KARNATAKA'S GOVERNANCE

Upon exploring the secondary data, we realized that the result was eye-opening. We could see the challenges and changes in agricultural practices between 1966 and 1980, which were reflected in the economic returns. Rainfall patterns, prices, availability of inputs, and even the presence of pests and diseases all played a role in shaping the instability in agricultural production. These fluctuations affected the area, production, and yield. Each crop showcased its own tale of unpredictability and challenges. Some crops saw a rise in instability, while others experienced a decline, leaving the farmers in a constant state of uncertainty. The plot thickened as we explored the district-level story and uncovered the intricate tale of Indian agriculture, where each state and district played a unique role. While the state-level data provided a broad overview of the challenges faced by farmers, the district-level analysis revealed the nuances and intricacies of instability at the grassroots level.

In Andhra Pradesh, there was a dynamic nature to crop cultivation spanning over five decades. However, the farmers faced challenges as diverse as the crops they grew. As the years passed, policies and climate conditions changed, and Andhra Pradesh showcased the intense uncertainties in the agricultural stage, where rice, cotton, groundnut, and chickpea took center stage (Chand and Raju, 2008). Paddy, groundnut, cotton, and chickpea saw significant fluctuations in the area, production, yield, fertilizer/pesticide consumption, and gross returns, which are evident across different periods (Table 2.1). There was a notable increase in both area and production from 1966–1980 to 2001–2017, suggesting agricultural expansion. However, this growth was accompanied by varying levels of the yield and fertilizer/pesticide consumption, reflecting changes in farming practices. Similarly, groundnut and cotton exhibit notable shifts in the production and yield over the years, indicating the influence of factors such as technological advancements, climate variability, and market dynamics. These fluctuations underscored the complex interplay of socio-economic and environmental factors shaping agricultural systems in Andhra Pradesh.

Meanwhile, in the neighbouring state of Telangana, similar tales unfolded with paddy, cotton, and maize. Looking at paddy cultivation, we noticed that the cultivated area, production, and income increased considerably over time, indicating its economic viability. However, the yield varied across the periods. The usage of fertilizers and pesticides also fluctuated. We observed similar trends for maize and cotton cultivation, with fluctuations in the area, production, yield, FHP, and gross returns across different periods (Table 2.2). These variations highlighted the dynamic nature of agricultural practices and economic factors influencing crop cultivation in the region over the years.

TABLE 2.1

Range of Instability (Coefficient of Variation %) of Crops in Andhra Pradesh between 1966 and 2017

		Area		Production		Yield		Farm Harvest Price		Gross Returns	
Crop	Period	District Level	State Level	District Level	State Level	District Level	State Level	District Level	State Level	District Level	State Level
Paddy	1966–1980	8.8–24.4	14.2	17.4–53.1	28.5	14.5–42	21.5	10.9–22.9	17.3	16.5–37.3	24.7
	1981–2000	6.7–43.8	19.2	13.8–59.8	31.4	10–31	17.3	1–20.4	12.1	12.8–32.7	19.6
	2001–2017	12.1–38.5	25.2	18.6–47.3	34.4	9–28.9	17.9	6.4–24.2	13.8	13.5–33.8	21.4
Groundnut	1966–1980	10.6–38.3	23.0	19.4–60.7	38.0	16.6–49.3	27.0	14–39.2	27.7	25.1–49.8	35.5
	1981–2000	10.3–25.3	16.4	16–66.3	34.5	14.5–60.4	28.2	12.7–40.7	18.9	13.6–66.3	32.4
	2001–2017	12–58.7	29.3	16.7–128.3	48.7	11.2–122	36.9	13–21.3	16.6	16–119.4	37.7
Cotton	1966–1980	0–98.7	39.8	0–121.1	49.5	0–82.9	37.6	17.7–40.6	21.5	42.1–95.1	67.2
	1981–2000	0–61.4	35.4	0–75.4	43.4	0–55.5	37.0	17.7–30	21.2	34.7–66.5	52.0
	2001–2017	19.5–72.3	40.2	29.2–82.3	55.0	24.6–76.5	40.3	15.5–27.3	20.2	24.5–76.4	43.3
Chickpea	1966–1980	0–66.3	37.4	0–78.3	32.3	0–84.9	33.8	13.1–31.7	15.8	32.2–94.2	55.3
	1981–2000	0–63.6	30.6	0–111	43.7	0–81.2	32.8	13.1–28.2	16.2	27.3–72.2	50.3
	2001–2017	0–106.7	47.8	0–102.8	61.1	0–100.6	34.8	6.3–16.4	13.1	20.3–96.2	39.0

TABLE 2.2
Range of Instability (Coefficient of Variation %) of Crops in Telangana between 1966 and 2017

		Area		Production		Yield		Farm Harvest Price		Gross Returns	
Crop	Period	District Level	State Level	District Level	State Level	District Level	State Level	District Level	State Level	District Level	State Level
Paddy	1966–1980	15.2–49.6	30.6–69.8	27.4–86.6	54.5	12.8–59.8	30	11.5–18.6	15.9	30.6–69.8	45.0
	1981–2000	16.9–41.2	16.8–38.3	24.8–58.2	40.3	10.1–39.8	20.1	8.4–23.9	13.5	16.8–38.3	22.7
	2001–2017	25.4–55.5	50.9–56.8	34.4–68.9	50.3	11.5–27.3	18.2	15.4–49.7	23	50.9–56.8	53.2
Maize	1966–1980	3.9–29.3	42.3–73.6	26.6–67.6	42.6	23–53	37	18.5–41.8	31.9	42.3–73.6	58.6
	1981–2000	3.9–45.6	28.1–60.9	16.1–58.0	32.7	12.4–47.7	28.5	14.8–25.5	21.1	28.1–60.9	41.2
	2001–2017	18.5–53.7	29.2–71.1	32–66.3	50.0	18.1–70.4	37.3	16.5–27.8	20	29.2–71.1	45.1
Cotton	1966–1980	7.9–140.2	52.4–109.6	0–104	39.5	0–88.6	32.6	24.4–55.9	49.9	52.4–109.6	74.1
	1981–2000	6.8–63.9	134.7–164.7	43.6–75.6	56.6	31.6–70.0	46.4	16.7–133	83.4	134.7–164.7	151.2
	2001–2017	14.9–37.5	76.7–94.6	23.7–60.5	42.2	20.5–46.3	36.8	17.9–72.2	25.6	76.7–94.6	84.9

TABLE 2.3

Range of Instability (Coefficient of Variation in %) of Crops in Karnataka between 1966 and 2017

		Area		Production		Yield		Farm Harvest Price		Gross Returns	
Crop	Period	District Level	State Level	District Level	State Level	District Level	State Level	District Level	State Level	District Level	State Level
Paddy	1966–1980	1.1–48.7	17.1	13.8–74	41.2	15.3–46.5	31.5	12.4–27.6	19.7	20.6–49.6	34.6
	1981–2000	1.8–57.1	20.0	10.1–73.6	35.4	6.9–58.1	23.7	9.4–25.8	15.3	13.9–57.4	26.0
	2001–2017	2–60.1	20.9	7.2–75.8	33.1	6.9–57.7	18.7	11.9–40.1	23.3	11.1–56.6	20.3
Sorghum	1966–1980	7.4–128.9	45.6	30.3–102.3	64.9	19.5–58.7	40.0	17–32.5	22.5	23.4–58.3	42.1
	1981–2000	6.7–55.8	27.1	24–85.3	47.7	17.3–64	34.5	11.6–39.1	26.1	27.5–69.7	44.3
	2001–2017	9.6–311.3	56.0	29.4–284.5	76.8	26.2–224.7	54.4	14.6–246.8	37.9	32.2–226.2	63.9
Pigeon pea	1966–1980	8.8–62.3	27.5	45.7–90.2	64.5	31.4–86.4	53.6	25.9–43.5	31.0	36–88.8	53.0
	1981–2000	8.2–65.7	30.8	39.6–98.5	67.0	31.5–95	58.2	18.5–193.3	31.1	38.6–98.7	61.1
	2001–2017	7.9–51.4	31.8	45.3–89.2	60.8	26.3–60.2	43.4	27.2–46	34.5	26.2–61.2	47.9

In Karnataka, farmers also encountered their share of hurdles. From the hilly sorghum farms to the endless paddy fields and pigeon peas, farming was not always smooth sailing. Across different periods, fluctuations were observed in various aspects of crop cultivation. We observed a noticeable increase in both area and production of paddy over the years, indicating an expansion of cultivation. However, the fluctuation in yield, fertilizer and pesticide usage, and gross returns reflected changes in farming practices (Table 2.3). Similarly, sorghum and pigeon pea cultivation also exhibit fluctuations in all aspects over time, highlighting the dynamic nature of agricultural practices.

It was evident that the farmers were facing a tough battle against the ever-changing nature of agriculture as the instability of crop yields had a profound impact on their lives. So, as researchers, it was our responsibility to do in-depth research using this data to come up with solutions and recommendations to help these hard-working farmers cope with these challenges.

REFERENCES

Chand R, Raju SS. (2008). Instability in Andhra Pradesh Agriculture—A Disaggregate Analysis. *Agricultural Economics Research Review*, 21(2), pp. 283–288. http://data.icrisat.org/district-level-data/

Dev MS. (1987). Growth and Instability in Food Grains Production: An Interstate Analysis. *Economic and Political Weekly*, 22(39), pp. A82–A92.

Hazell PBR. (1982). *Instability in Indian Food Grain Production*. Research Report, 30, International Food Policy Research Institute, Washington, DC.

Khatri P, Kumar P, Shakya KS, Kirlas MC, Tiwari KK. (2024). Understanding the Intertwined Nature of Rising Multiple Risks in Modern Agriculture and Food System. *Environment, Development and Sustainability*, 26(9), pp. 24107–24150.

Reddy AA. (2006). Growth and Instability in Chickpea Production in India: A State Level Analysis. *Agricultural Situation in India*, 63, pp. 130–145.

Reddy AA. (2012). Sources of Agricultural Productivity Growth in Orissa, India: A Regional Perspective. *Regional Development Studies*, 16, pp. 99–116.

Senapatia AK, Goyari P. (2019). Growth and Instability in Agricultural Productivity in Odisha. *Agricultural Economics Research Review*, 32(1), pp. 55–65.

Sharma HR, Singh K, Kumari S. (2006). Extent and Source of Instability in Food Grains Production in India. *Indian Journal of Agricultural Economics*, 61(4), pp. 647–666.

Sravani N. (2023). Erratic rain pattern makes farming an unsustainable future in Andhra Pradesh. The Hindu. July 28, 2023. https://www.thehindu.com/news/national/andhra-pradesh/farming-an-unsustainable-future/article67123842.ece

3 Strategies for Gathering Evidence

THOUGHT OF A SURVEY

Once a farmer said, "*Climate change means there are now seasons when the harvest fails. If this happens the immediate impact is hunger for the already vulnerable families.*"

The 2021 outbreak of *Thrips parvispinus* (Kudo) in the chilli fields of Telangana, Andhra Pradesh, and Karnataka has generated significant concern among farmers and agricultural stakeholders (Kumar et al., 2024). This cosmopolitan species, recognized for its quarantine importance, was initially reported in regions ranging from Thailand to Australia (Mound & Marullo, 1996). Over the past two decades, *T. parvispinus* has significantly expanded its geographic range, now being recorded in countries such as France, Greece, Hawaii, Mauritius, Reunion, Spain, Tanzania, and the Netherlands, as well as India (Rachana et al., 2018).

T. parvispinus is highly polyphagous, inflicting considerable damage on various crops (Morse & Hoddle, 2006). Additionally, it poses a threat to ornamental plants such as anthurium, chrysanthemum, dahlia, dipladenia, gardenia, and ficus (CABI, 2020). In India, this species was first reported on papaya in Bengaluru, Karnataka, in 2015 (Tyagi et al., 2015). However, since October 2021, it has become a major pest in chilli cultivation, causing severe damage to fields (ICAR, 2021). This proved to be true when chilli farmers in Telangana, Andhra Pradesh, Karnataka, and some other states started living their worst nightmare. Climate change, coupled with a new infestation, destroyed hectare after hectare of standing crops in a matter of months (Shagun, 2022) (Figure 3.1).

The damage ranged from 70% to 100% in the year 2021 (Veranna et al., 2022), especially in Telangana, Andhra Pradesh, and Karnataka, where many farmers lost 100% of their chilli crops, with losses amounting to more than one lakh per acre and leaving the farmers in a severe financial crisis. Our secondary research already sheds some light on the unpredictable world of agriculture and the resilience of the people who depend on it. As researchers, we were curious about this fact, and to get the experience of witnessing the harsh realities in these states, the idea of conducting a household survey came to our minds. Our team consisted of a group of researchers, including myself, and we thought to embark on a journey to unravel the secrets behind the farmers' resilience.

HOW DID WE PLAN IT?

Based on the information we collected through the secondary research, we prepared action plans and a structured questionnaire, which helped us address the issues. We

DOI: 10.1201/9781003582878-3

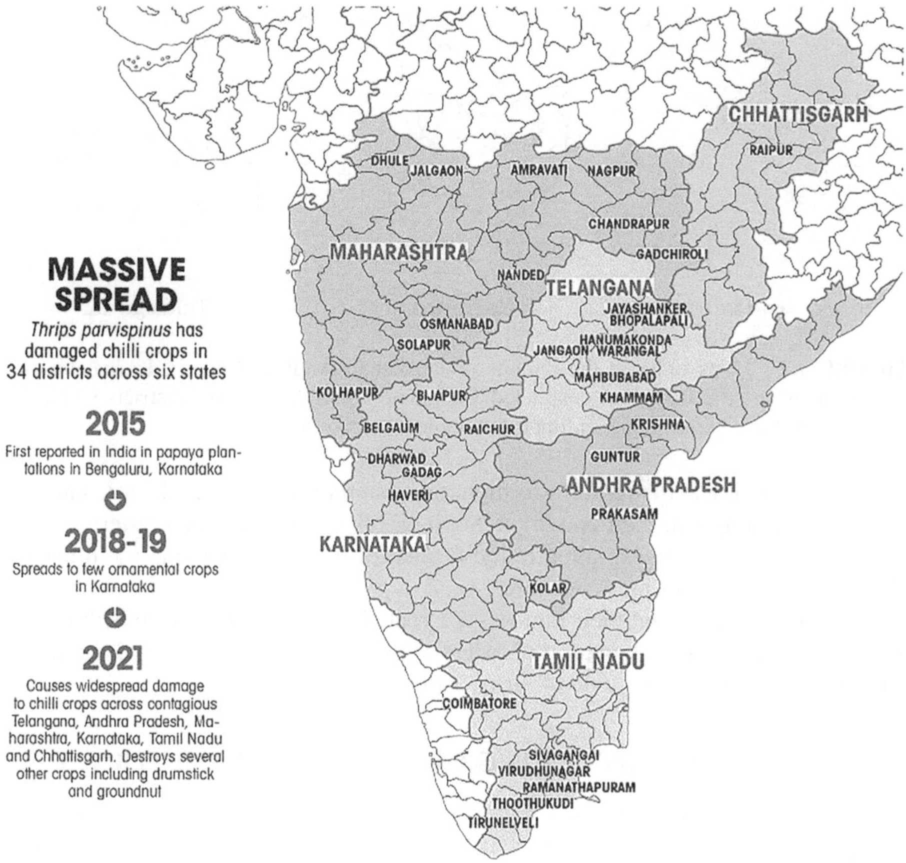

FIGURE 3.1 Map of southern India, presenting the spread of the pest in chilli crops across 34 districts in Maharashtra, Telangana, Andhra Pradesh, Karnataka, Tamil Nadu, and Chhattisgarh. Though the pest was first reported in Karnataka in 2015, it caused widespread damage to chilli crops in contiguous states in 2021.

started preparing ourselves in such a way that if farmers asked us questions about the survey, we answered them with appropriate figures. We felt that we should collect detailed information to capture a very clear picture.

Our investigation led us to three states, each with its unique way of handling agricultural risks. While Andhra Pradesh introduced the YSR Free Crop Insurance Scheme and Karnataka joined the Pradhan Mantri Fasal Bima Yojana, Telangana remained without a safety net (Reddy et al., 2024). We delved into the depth of these policies to understand their impact on farmer welfare. We randomly picked three districts from each state and then selected two blocks from each district. In four villages per block, we met 20 farmers each – a total of 1,440 farmers who became our storytellers. Figure 3.2 indicates the study area blocks in three states.

Our goal? Understand the levels of farmers' distress by peeling the back layers to reveal a gripping story. The farmers, who worked hard day and night to feed the

FIGURE 3.2 Maps indicate the regions of interest in Andhra Pradesh, Karnataka, and Telangana where the study was conducted. Each map highlights specific districts within the states, with shaded areas representing the study locations or blocks.

nation, were living in distress. It was high time that their voices were heard, and steps were taken to improve their condition. So, it was like exploring the mystery of challenges, resilience, and adaptation in the farmers' world and assessing the impact of different policies on farmers' welfare.

The more farmers and their families we visited, the more stories we heard. From farmers being affected by natural disasters to those exploited by money lenders, our survey revealed the tragic state of farmers in our country.

CHALLENGES OF CONVINCING FARMERS ABOUT THE SURVEY

During the time of the survey, we prepared the timetable and shared responsibilities by identifying who would survey different villages. Armed with questionnaires, when we ventured into the villages, we were met with open arms by the farmers and were taken in by their warm hospitality. We developed a good relationship with the farmers in the villages and inquired about their village life and well-being. But a little problem arose once we went house to house to do the survey.

"What is a survey?"

"Why are you doing the survey?"

"How will it benefit us?"

These questions started coming from the farmers and their families. Since we were well prepared for our aim, we were able to convince people about the survey. After that, they trusted us and talked about the situation of their farms and households. Gradually we dove deeper into our research and into their lives during the 2020 crop year, capturing tales of cost, cultivation, and resilience, and we were stunned by what we discovered. Their voices became the heartbeats of our journey.

WHAT NEXT AFTER DATA COLLECTION?

We developed a tool called the Farmers Distress Index (FDI), considering factors like clarity, relevance, and ease of data collection. It was a tool not just to identify distress but to understand its depth and form. A comprehensive look at the farmer's distress, decomposed into seven dimensions, allows us to map vulnerabilities at the sub-district level. As we reached the end of our research, we put our FDI through a robustness check, a validation across geographies and farmer groups. And so, the journey concludes, echoing the resilience of farmers, the complexity of policies, and the power of a simple yet robust tool – FDI. Our research was just a small step towards unfolding the misery of farmers. It was an adventure that opened our eyes to the harsh realities faced by these simple and hardworking individuals.

REFERENCES

CABI. (2020). *Thrips parvispinus*. Invasive Species Compendium.

ICAR. (2021). *Report on Thrips parvispinus Infestation in Chilli*. Indian Council of Agricultural Research.

Kumar PS, Rachana RR, Amarendra B, Sheela N. (2024). Transient Composition of the Thrips Species (Thysanoptera: Thripidae) Infesting Mulberry in Southern India: First Report of Two, Including the Dominating Invasive Pest *Thrips parvispinus*. *Journal of Integrated Pest Management*, 15(1), p. 32. https://doi.org/10.1093/jipm/pmae024

Morse JG, Hoddle MS. (2006). Invasion Biology of Thrips. *Annual Review of Entomology*, 51 (1), pp. 67–89.

Mound LA, Marullo R. (1996). *The Thrips of Central and South America: An Introduction (Insecta: Thysanoptera)*. Associated Publishers.

Rachana RR., Roselin P, Varatharajan R. (2018). Report of Invasive Thrips Species, Thrips *parvispinus* (Karny)(Thripidae: Thysanoptera) on *Dahlia rosea* (Asteraceae) in Karnataka. *Pest Management in Horticultural Ecosystems*, 24(2), pp. 175–176..

Reddy AA, Rani R, Das A, Reddy CP, Sneha AV. (2024). Mapping Welfare and Development Schemes to SDGs at Village Level in India. *SAARC Journal of Social Science*, 2(1), pp. 27–59.

Shagun. (2022). Chilli under attack: New thrips species destroys farms in 6 states. Down to Earth. https://www.downtoearth.org.in/news/agriculture/chilli-under-attack-new-thrips-species-destroys-farms-in-6-states-82811

Tyagi K, Kumar V, Singha D, Chakraborty R. (2015). Morphological and DNA Barcoding Evidence for Invasive Pest Thrips, Thrips parvispinus (Thripidae: Thysanoptera), Newly Recorded from India. *Journal of Insect Science*, 15(1), p. 105.

Veeranna D, Reddy RU, Moguloju M, Padmaja G. (2022). Report on Heavy Infestation and Damage by Invasive Thrips Species, *Thrips parvispinus* (Karny) on Chilli in Telangana State of India. *The Pharma Innovation Journal*, 11(7), pp. 3845–3848.

4 Farmer Assets and Wealth Dynamics

"*We spend more on cultivation and there is no earning,*" said a farmer. "*I hardly harvested enough,*" added another farmer.

As we drove through the villages, we heard the shouts of farmers. We saw small plots of land scattered across the fields. Some were as small as 2 hectares, while others were larger, spanning over 10 hectares. On average, about 1.08 hectares (Agriculture Census Division, 2018) were the livelihoods of farmers, who looked after them. We knew that, despite the size of these plots of land, they put in their hard work to make the most of what they had because their livelihoods depended on the success of their crops. There was an endless variety of crops. From rice to wheat, pulses, fruits, and vegetables, what was not there? But the earnings were severely declining (Vasavi, 1999; Mohanty, 2005; Deshpande, 2002; Sidhu et al., 2011; Shah, 2012), and after investing a lot they have not been able to cope with the rising costs of living (Desai, 2016; Patnaik, 2024). So they practised subsistence farming, selling some of their produce in local markets and counting on government support to make ends meet. Farmers used different ways to make it safer and to adjust themselves to the risks and changes (Stringer et al., 2020). In the field, some farmers used age-old knowledge passed down from their ancestors while others embraced modern techniques and tools to take care of their fields.

But life is not easy for these farmers. One farmer said with regret:

> We are facing severe water shortages for growing crops. We are in deep stress.

As researchers, we witnessed the farmers' struggles earlier. Every year, the monsoon was unpredictable – sometimes it flooded the fields, and sometimes left with drought. This constant uncertainty made farming incredibly difficult for the farmers.

They battled against many unpredictable challenges along the way. They also struggled to access important resources like seeds, fertilizers, credit, markets, and government support (Mohanty and Lenka, 2019). Besides, there was bad weather that brought uncertainties, like droughts and floods, and devastated their crops and livelihoods. So the living standards of families were quite complicated to understand because there were lots of different things to know. The main fact was that some people did not get equal chances of getting what they needed to grow (Kuang et al., 2020; Udoh et al., 2017; Jansen et al., 2006; Yang et al., 2018; Lindenberg, 2002; Reddy et al., 2014). While some farmers engaged solely in agriculture, others explored other avenues beyond agriculture by venturing into non-agricultural activities and self-employment.

But there lies something as the hidden treasure in the heart of every village: farmer's physical assets. These assets were like magic tools, helping them make their fields flourish by increasing production and helping the farmers get out of their misery (Yang et al., 2018; Reddy, 2021). It does everything from planting seeds to delivery

 DOI: 10.1201/9781003582878-4

to the market. From the ponds that provided water for irrigation to the tractors that ploughed the land, every asset had a crucial role to play in the farmers' lives. But there was something more that caught our attention. Farmers also had some reliable companions – their livestock. The livestock made their work lighter and the harvest richer, and it was expected to help them improve their situation (Brown, 2003; Catley, 2008; Delgado, 2003; ILRI, 2007, 2011; Pica-Ciamarra, 2009; Sekaran et al., 2021). As we spent time with the farmers, we realized that their bond with their livestock was unbreakable. They treated them with love and respect, knowing that they were an essential part of their livelihood.

However, these assets were not just limited to agriculture. In the village, assets like TVs, refrigerators, smartphones, vehicles, and LPG were not only part of everyday life but also indicated the advanced infrastructure (Sedai et al., 2021). These little things made life in the village happy and cosy. It is like having a helping hand with everyday tasks and making life much better and easier in rural areas (Vu et al., 2020). Hence these little luxuries may seem small, but they were symbols of hope, prosperity, and progress in the village.

Out of the total sample of 1,440 households in three states, namely Andhra Pradesh, Karnataka, and Telangana, the majority in all three states are Hindu (94.7%) households by religion, followed by Muslims (3.7%) and Christians (1.6%) (Table 4.1). The share of Hindus is higher in Telangana (97.2%), followed by Karnataka (94%) and Telangana (93.1%). The share of Hindu households is higher among the sample households as the sample is collected from mainly those who are directly dependent on agriculture as a major source of income and who claim themselves as farmers. In general, the share of farmers in the Hindu religion is higher than both Muslims and Christians as the latter group's engagement in non-agricultural activities and self-employment in non-agriculture are higher. Overall, in the sample share of forward-caste (FC) households is higher at 40.9%, followed by other backward castes (OBCs) with 39.6% and scheduled tribe (ST, 10.3%), and the least is for scheduled caste (SC, 9.2%). Probably, the sample is more biased towards landowners; hence, the share of FCs and STs was higher than their actual share in the population, but they reflect that, when compared to their share in the population, the representation of FCs and STs in farming activities as cultivators is more. Again, there is some underrepresentation of tenant-farmers in the sample, although we try to incorporate tenant-farmers in the sample, as many tenant-farmers have not come forward for answering to our questionnaire probe due to the fear of landowners.

There are some slight inter-state differences in the share of social groups: the share of FCs was higher in Andhra Pradesh, while the share of OBCs was higher in Telangana and the share of STs was higher in Karnataka. The share of pakka houses in the sample households is 76.8%, highest in Andhra Pradesh (89.6%), followed by Telangana (73.9%) and Karnataka (70.3%). The reported tenancy was 5.1% in the overall sample, highest at 7.7% in Andhra Pradesh, followed by Karnataka (5.8%) and least in Telangana (1.5%). However, this reported tenancy is an underestimation of the prevalence of actual tenancy due to many reasons like fear from landowners not to disclose openly, the informal nature of tenancy which is not legal in any respect, and no written document to authenticate the tenancy. As per the poverty line, poverty levels in the study area are 33.2%, highest in Andhra Pradesh (39.1%) and lowest in

TABLE 4.1
Basic Socio-Economic Status of Sample Households (% of Households)

	States			
	Andhra Pradesh	Karnataka	Telangana	All
Religion				
Hindu	93.1	94.0	97.2	94.7
Muslim	2.7	6.0	2.2	3.7
Christian	4.2	0.0	0.7	1.6
Social Groups				
SC	9.8	11.5	6.3	9.2
ST	2.5	13.5	15.0	10.3
OBC	25.6	37.9	55.9	39.6
FC	62.1	37.1	22.8	40.9
House Type				
Katcha	10.2	29.3	26.1	23.1
Pakka	89.6	70.3	73.9	76.8
Tenancy				
Owner	92.3	94.2	98.5	94.9
Tenant	7.7	5.8	1.5	5.1
Poverty Status (2020–2021)				
Non-poor	57.7	64.8	78.3	66.8
Poor	42.3	35.2	21.7	33.2
Land Size				
Marginal (<1 acre)	18.8	10.4	12.6	13.9
Small (1–2 acre)	17.1	14.8	21.7	17.8
Medium (2–4 acre)	29.6	27.9	39.1	32.1
Semi-large (4–10 acre)	28.3	34.0	24.6	29.0
Large (>10 acre)	6.2	12.9	2.0	7.1
All	100.0	100.0	100.0	100.0

Note: % indicates the proportion of households within each category.

Telangana (21.7%). However, these poverty numbers should not be generalized to the whole population of the respective states, as these districts are selected purposively and do not represent whole states. We follow the land size classification, followed by the Indian government with less than 1 acre as marginal farmers, 1–2 acres as small farmers, 2–4 acres as medium farmers, 4–10 acres as semi-large, and more than 10 acres as large farmers. Marginal landholding farmers are more, and large landholders are less in Andhra Pradesh as against more large farmers and less marginal farmers in Karnataka. Overall, 13.9% are marginal landholders, 17.8% are small landholders, 32.1% are medium landholders, 29.0% are semi-large, and 7.1% are large landholders.

Among the states, as per the land capabilities, Andhra Pradesh lands are more fertile and can support more population per unit area, while Karnataka lands are less fertile and can support less population per unit area; the results seem to reflect the land capability to support households. Our field observations also reflected these inter-state differences, however with some changes that happened over the past two decades.

Our survey district in Andhra Pradesh is around Amaravati (Amaravati was declared as a state capital recently), where land values have increased recently, but there is no commensurate development and non-farm opportunities for employment and income. In Andhra Pradesh, in some villages farmers are stuck up in old cropping patterns like rice cultivation, which is not as profitable as that of commercial crops like cotton and chillies. On the contrary, farmers in Telangana have shifted to high-risk–high profitable cropping patterns like chillies and cotton; as a result, their incomes from agriculture have increased but with huge risk. In northern Karnataka, farmers are still cultivating paddy and other diversified crops like jowar, millets, pulses, and oilseeds, but their level of development in all respects is lower when compared to survey districts of Andhra Pradesh and Telangana. Their level of mechanization and commercialization is low in terms of tractorization and use of inputs like fertilizers and pesticides and cultivated areas under commercial crops like cotton and chillies. Overall the selected sample households cannot be interpreted as a representative sample of the respective states. Instead, they serve as case studies of farmers who experienced high levels of hazard in 2021, with over 50% reporting crop losses due to thrips attacks on chilies. These farmers' experiences are contrasted with a relatively normal agricultural year in 2020. The design of the sample and selection of the sample households are purposeful, mainly to assess coping and adaptation strategies for exposure to the high hazard in the crop year 2021 when compared to the normal year 2020.

WHO OWNS WHAT? EXPLORING WEALTH DISTRIBUTION

Religion Matters

As we dug deeper into each group of farmers' community, we noticed a stark difference in the size of the land. All the Hindu and Muslim families we talked to owned big pieces of land, about 5 acres each. But the land of the Christian people we met was smaller, around 2.4 acres on average. Hindus seemed to have more land-related assets compared to Christians (Table 4.2). Not many households had livestock, especially animals like sheep, goats, and chickens, except for cows. Hindus seemed to have more cows compared to others. This could be because they depend a lot on farming for their livelihood. Not only that, about 15% of households owned tractors, and more Hindus seemed to have them. They also had more farming machines compared to the other religions. This could be because they have a higher income and can afford to invest in these machines. Even though Hindus were more advanced, it was heartwarming to see that most households had TVs, smartphones, two-wheeler vehicles, and LPG, and some even had fridges.

Also, there were vast variations in wealth among different religious groups (Table 4.3). Hindu households had higher residential house values and more cultivated land compared to Christians. On average, Hindu families had total assets

TABLE 4.2
Physical Assets Available with Households by Religious Groups

	Religion							
	Hindu		Muslim		Christian		Total	
Number of farmers	1,345		52		23		1,420	
	Mean	%	Mean	%	Mean	%	Mean	%
Land-Related Assets								
Cultivated land (acres)	5.0	96	5.0	90	2.4	87	5.0	96
Cattle shed/farmhouse (no.)	1.0	18	1.0	8	1.0	13	1.0	18
Farm pond (no)	1.0	6	1.0	6		0	1.0	6
Storage house (no)	1.1	2		0		0		0
Livestock								
Working cattle/buffalo (agriculture)	2.0	21	2.2	8	4.5	9	2.0	20
Milk cattle/buffalo	1.9	36	2.2	23	1.8	22	1.9	35
Young stock	1.6	15	1.3	6	2.0	17	1.6	15
Sheep	10.4	5	3.7	6		0	10.1	5
Goats	8.0	2		0		0	8.0	2
Poultry	8.1	9	3.8	8	3.7	13	7.9	9
Agricultural Assets (Equipment and Machinery)								
Tractor/trolley	1.0	16	1.2	10	1.0	4	1.0	15
Power tiller	1.0	5		0	1.0	4	1.0	4
Bullock cart	1.0	9	1.0	4	1.0	4	1.0	9
Wooden/iron plough	1.0	25	1.0	27	1.0	13	1.0	24
Sprayer/duster	1.0	58	1.0	48	1.0	26	1.0	57
Weeder	1.0	14	1.0	17	1.0	9	1.0	14
Seed drill/hoes	1.0	8	1.0	6	1.0	9	1.0	8
Infrastructure (Transportation and Communication)								
Bicycle (non-agriculture)	1.0	12	1.0	15	1.0	13	1.0	13
Auto rickshaw	1.0	3	1.0	13		0	1.0	3
Two wheelers	1.0	77	1.1	73	1.0	57	1.0	77
Four wheelers	1.0	3	1.0	4		0	1.0	3
Smartphone	1.1	78	1.1	69	1.0	43	1.1	77
Basic phone	1.0	66	1.1	65	1.0	91	1.0	66
Non-farm Asset (Lifestyle Goods)								
Sewing machine	1.0	5	1.0	13	1.0	4	1.0	5
TV	1.0	89	1.0	87	1.0	87	1.0	89
Computer/laptop	1.1	6	1.0	4	1.0	9	1.1	6
Refrigerator	1.0	43	1.0	31	1.0	30	1.0	43
LPG with cooking set	1.0	93	1.0	94	1.0	96	1.0	93

Note: % indicates the proportion of households within each category.

valued at about Rs. 64 lakhs, while Muslim households had about Rs. 46 lakhs and Christian households had about Rs. 29 lakhs.

TABLE 4.3
Religion-Wise Distribution of Household Assets' Value

	Religion						Total	
	Hindu		Muslim		Christian			
	Mean	%	Mean	%	Mean	%	Mean	%
Residential house present value (Rs.)	6,10,935	99	5,00,980	98	4,06,818	96	6,03,743	99
Cultivated land present value (Rs.)	59,38,180	96	44,82,660	90	27,22,500	87	58,40,446	96
Cattle shed/ farmhouse present value (Rs.)	51,989	17	34,833	6	41,667	13	51,641	17
Farm pond present value (Rs.)	77,183	5	47,000	6		0	75,959	5
Storage house present value (Rs.)	2,29,765	1		0		0	2,29,765	1
Total assets value (Rs.)	63,57,354	100	46,15,010	100	28,94,565	100	62,37,462	100

Note: % indicates the proportion of households within each category.

Social Strata

Surprisingly, the historical influence of the caste system in India still affects who owns what (Rao, 2017). On average, families had about 5 acres of land. Richer families, especially the FC ones, had bigger plots, while SC families had smaller ones (Table 4.4). Even though the government was trying hard to help, a lot of inequality still existed. It was interesting to see that SC families owned fewer assets, except for livestock like sheep and goats. They even had more tractors than others. It was also good that most families, irrespective of groups, had phones and LPG. Looking closer, we saw that some groups had more valuable homes or land and owned more than double the wealth of others (Table 4.5). This made us think about how fair things are and raised a question if we are doing enough to make things equal for everyone.

Land Conflict

In the land of marginal farmers, where the plots were modest at 1.3 acres, we noticed a large number of small ruminants like sheep and goats as the substitute for their limited land and income for making the most (Table 4.6). In contrast, since the large farmers could afford bigger investments, they owned more cattle and poultry. But we

TABLE 4.4
Physical Assets Available with Households by Social Group

	Caste								Total	
	SC		ST		OBC		FC			
Number of farmers	131		146		562		581		1,420	
	Mean	**%**	**Mean**	**%**	**Mean**	**%**	**Mean**	**%**	**Mean**	**%**
Land-Related Assets										
Cultivated land (acres)	3.1	91	4.0	97	5.0	95	5.7	97	5.0	96
Cattle shed/farmhouse (no.)	1	8	1.2	21	1.0	19	1.0	18	1.0	18
Farm pond (no)	1	7	1.0	12	1.0	6	1.0	4	1.0	6
Storage house (no)	1	2	1	1	1	2	1.2	1	1.1	1
Livestock										
Working cattle/buffalo (agriculture)	2.4	12	1.9	36	1.9	23	2.1	15	2.0	20
Milk cattle/buffalo	1.8	24	1.7	34	1.8	38	2.2	36	1.9	35
Young stock	1.3	8	1.4	7	1.6	17	1.6	15.	1.6	15
Sheep	6.6	5	13.0	14	10.5	5	4.0	1	10.1	5
Goats	4.5	3	2.5	7	12.8	3	2.0	0	8.0	2
Poultry	9.14	11	9.9	24	7.0	9	6.4	5	7.9	9
Agricultural Assets (Equipment and Machinery)										
Tractor/trolley	1.1	9	1.0	13	1.0	12	1.0	20	1.0	15
Power tiller	1	2	1.0	5	1.0	2	1.1	7	1.0	4
Bullock cart	1	5	1.0	11	1.0	12	1.0	7	1.0	9
Wooden/iron plough	1	18	1.0	26	1.0	30	1.0	20	1.0	24
Sprayer/duster	1.0	40	1.0	53	1.0	61	1.0	58	1.0	57
Weeder	1	6	1.0	12	1.0	19	1.0	12	1.0	14
Seed drill/hoes	1	6	1.0	5	1.0	8	1.0	10	1.0	8
Infrastructure (Transportation and Communication)										
Bicycle (non-agriculture)	1	17	1.0	16	1.0	16	1.0	7	1.0	13
Auto rickshaw	1	2	1.0	5	1.0	4	1.0	2	1.0	3
Two wheelers	1.0	57	1.0	65	1.0	81	1.1	80	1.0	77
Four wheelers	1	1	1.0	2	1.0	4	1.0	4	1.0	3
Smartphone	1.1	71	1.2	73	1.1	75	1.2	81	1.1	77
Basic phone	1.0	68	1.0	64	1.0	63	1.0	70	1.0	66
Non-farm Assets (Lifestyle Goods)										
Sewing machine	1	5	1.0	5	1.0	7	1.0	4	1.0	5
TV	1	85	1.0	79	1.0	90	1.0	92	1.0	89
Computer/laptop	1.33	2	1.0	3	1.0	6	1.1	7	1.1	6
Refrigerator	1	21	1.0	34	1.0	43	1.0	49	1.0	43
LPG with cooking set	1	91	1.0	92	1.0	93	1.0	95	1.0	93

Note: % indicates the proportion of households within each category.

TABLE 4.5

Social Group-Wise Distribution of Household Asset Values

	Caste									
	SC		ST		OBC		FC		Total	
	Mean	%	Mean	%	Mean	%	Mean	%	Mean	%
Residential house value (Rs.)	4,32,326	98	5,80,308	100	6,13,664	99	6,38,591	99	6,03,743	99
Number of rooms (no)	3	98	3	100	3	99	3	99	3	99
Cultivated land value (Rs.)	28,15,042	91	34,90,352	97	67,20,244	95	62,39,034	97	58,40,446	96
Cattle shed/farmhouse value (Rs.)	28,182	8	70,846	18	54,250	18	46,620	17	51,641	17
Farm pond value (Rs.)	41,625	6	72,000	10	54,778	5	1,12,040	4	75,959	5
Storage house value (Rs.)	11,667	2	1,00,000	1	3,21,375	1	2,40,000	1	2,29,765	1
Total assets value (Rs.)	30,19,756	100	39,95,240	100	70,16,343	100	67,73,009	100	62,37,462	100

Note: % indicates the proportion of households within each category.

TABLE 4.6
Physical Assets Available with Households by Cultivated Land Category

	Total Cultivated Land-2022									
	Marginal		Small		Semi-Medium		Medium		Large	
Number of farmers	80		195		407		506		232	
	Mean	%	**Mean**	%	**Mean**	%	**Mean**	%	**Mean**	%
Land-Related Assets										
Cultivated land (acres)	1.3	99	1.9	95	3.1	96	5.4	95	11.5	95
Cattle shed/farmhouse (no.)	1.0	8	1.2	9	1.0	14	1.0	21	1.0	29
Farm pond (no)	1.0	6	1.0	4	1.0	7	1.0	7	1.0	5
Storage house (no)	1	3	1	1	1	2	1	1	1	2
Livestock										
Working cattle/buffalo (agriculture)	2	8	2	12	2	18	2	22	2	28
Milk cattle/buffalo	2	25	2	17	2	33	2	38	3	51
Young stock	1.9	9	1.4	5	1.3	12	1.5	17	1.8	24
Sheep	25.8	5	15.2	7	5.6	4	8.3	5	8.1	4
Goats	2.0	1	1.9	4	8.6	2	12.1	2	7.3	2
Poultry	6.3	5	5.8	12	7.4	10	8.9	8	10.1	7
Agricultural Assets (Equipment and Machinery)										
Tractor/trolley	1.0	1	1.0	6	1.1	5	1.0	14	1.0	48
Power tiller	1.0	1	1.0	1	1.0	2	1.0	3	1.1	15
Bullock cart	1.0	3	1.1	5	1.0	6	1.0	10	1.0	17
Wooden/iron plough	1.0	11	1.0	14	1.0	19	1.0	25	1.0	47
Sprayer/duster	1.0	33	1.0	39	1.0	45	1.0	67	1.0	78
Weeder	1.0	4	1.0	7	1.0	8	1.0	15	1.0	33
m. Seed drill/hoes	1.0	5	1.0	5	1.1	4	1.0	9	1.0	17
Infrastructure (Transportation and Communication)										
Bicycle (non-agriculture)	1.0	13	1.0	16	1.0	16	1.0	11	1.1	6
Auto rickshaw	1.0	6	1.0	3	1.0	3	1.0	3	1.0	3
Two wheelers	1.0	60	1.0	64	1.0	69	1.0	83	1.1	93
Four wheelers	1.0	3	1.0	1	1.1	2	1.0	3	1.0	10
Smartphone	1.1	71	1.1	63	1.1	70	1.1	83	1.3	90
Basic phone										
Non-Farm Assets (Lifestyle Goods)										
Sewing machine	1.0	6	1.0	4	1.0	5	1.0	6	1.0	8
TV	1.0	89	1.0	88	1.0	86	1.0	91	1.0	93
Computer/laptop	1.0	6	1.0	3	1.1	4	1.0	6	1.1	10
Refrigerator	1.0	34	1.0	25	1.0	37	1.0	46	1.0	63
LPG with cooking set	1.0	93	1.0	89	1.0	93	1.0	94	1.0	97

Note: % indicates the proportion of households within each category.

were really amazed to find a harmonious balance of essential agricultural infrastructure. These showed a shared commitment to nurturing the land and utilizing it to its full potential. It was also nice to see having various lifestyle goods across all the categories to improve the living standard and household comfort.

The total asset value of the farmers varied a lot between 18 lakhs to 2.5 crores (Table 4.7), which questions the expansion of the public–private system for the betterment of smaller farmers (Kadaraiah et al., 2022; Mehta et al., 2019; Sarkar, 2020; Aryal et al., 2019, 2021). But the thing that surprised us most was that most of the asset value came from land and houses. It seemed the large farmers mostly relied on their land and houses for their livelihoods.

Poverty Status Matters

"*No land, no money, how to buy a tool?*" A tone of pain in an old farmer's voice when we asked about what assets they had.

Though we were a little embarrassed with such an answer, we realized it was true. If one-third of the farmers were below the poverty line, which jumped to 47% after the crop failure, what assets can these people be expected to have?

Nevertheless, we continued our conversation, to know their exact position. It was obvious to find that non-poor households owned 1.5 times more land than poor households (Table 4.8). Even though they had a good number of small ruminants like goats and sheep, they lost many after the crop failure. Except for the more usage of smartphones in non-poor households, the asset conditions of both of these groups were quite similar. Naturally, the non-poor households were quite wealthier than the poor households. But we were a little surprised when we saw that crop loss did not impact much on the wealth of non-poor households. The asset's value remained the same after the disaster (Table 4.9). Even more surprising was the fact that, among the poor households, the asset values had increased by about 13%.

Tenancy Impact

Something informal that made us very curious was that some farmers called tenants held exclusive usage rights over the land but without ownership, as the wealthy landlord owned it and maintained full control.

> "*It is an uncertain position to be in, but it is the only way for us to survive. We have no other options*," one helpless farmer cried suddenly.

We were shocked, but we proceeded to our work. We made these two types of households a separate group in the farming community. The asset difference was a bit intriguing. Our hearts sank when we discovered that a staggering 82% of the tenant-farmers did not own any land at all (Table 4.10). They relied more on traditional equipment, while owner-farmers had modern machinery tenants. Also, the tenant-farmers said a similar thing we found in some other studies as well that they sometimes move in and out to get employment in non-agricultural income sources instead of solely relying on agriculture (Bansal, 2020; Bhattacharya, 2019; Baruah et al., 2022; Hossain et al., 2019). It was a harsh reality that the more cultivable land of the owner-farmers amounted to a wealth value of Rs. 65 lakh, which was five times that of the tenant-farmers (Table 4.11).

TABLE 4.7
Asset Value by Household Land Size Categories

	Marginal		Small		Semi-Medium		Medium		Large		Total	
Variable	**Mean**	%	**Mean**	%	**Mean**	%	**Mean**	%	**Mean**	%	**Mean**	%
Residential house value (Rs.)	5,77,026	98	5,14,683	99	5,80,586	99	6,40,371	99	8,34,653	100	6,03,743	99
Number of rooms (no)	3	98	3	99	3	99	3	99	4	100	3	99
Cultivated land value (Rs.)	17,10,357	70	21,96,059	100	38,92,615	100	71,24,447	100	2,43,67,228	100	58,40,446	96
Cattle shed/farmhouse value (Rs.)	37,308	7	65,346	10	44,083	17	55,761	23	52,885	26	51,641	17
Farm pond value (Rs.)	37,857	4	86,111	4	67,773	5	89,355	8	64,000	5	75,959	5
Storage house present value (Rs.)	52,500	1	10,07,500	1	66,200	1	1,38,333	1	3,12,500	2	2,29,765	1
Total assets value (Rs.)	18,08,920	100	27,73,075	100	45,01,697	100	78,04,170	100	2,52,34,752	100	62,37,462	100

Note: % indicates the proportion of households within each category.

TABLE 4.8
Ownership of Assets by Poverty Status

	2020–2021						2021–2022					
	Non-Poor		Poor		Total		Non-Poor		Poor		Total	
Number of farmers	948		472		1,420		0	755	665		1,420	
	Mean	%	Mean	%	Mean	%	Mean	%	Mean	%	Mean	%
Land-Related Assets												
Cultivated land (acres)	5.6	95	3.9	96	5.0	96	5.6	95	4.3	96	5.0	96
Cattle shed/farmhouse (no.)	1.0	20	1.0	14	1.0	18	1.0	19	1.0	16	1.0	18
Farm pond (no)	1.0	6	1.0	6	1.0	6	1.0	5	1.0	7	1.0	6
Storage house (no)	1	1	1	2	1	1	1	1	1	2	1	1
Livestock												
Working cattle/buffalo (agriculture)	2	22	2	17	2	20	2	22	2	17	2	20
Milk cattle/buffalo	2	39	2	32	2	35	2	39	2	32	2	35
Young stock	1.6	15	1.5	13	1.6	15	1.7	15	1.4	14	1.6	15
Sheep	11.4	5	6.3	4	10.1	5	12.8	5	5.5	4	10.1	5
Goats	6.7	2	10.6	2	8.0	2	6.0	2	9.7	3	8.0	2
Poultry	8.0	10	7.8	8	7.9	9	8.1	9	7.7	9	7.9	9
Agricultural Assets (Farm Machinery and Equipment)												
Tractor/trolley	1.0	18	1.1	9	1.0	15	1.0	19	1.0	11	1.0	15
Power tiller	1.0	6	1.0	1	1.0	4	1.0	6	1.0	2	1.0	4
Bullock cart	1.0	11	1.0	5	1.0	9	1.0	10	1.0	7	1.0	9
Wooden/iron plough	1.0	29	1.0	15	1.0	24	1.0	30	1.0	19	1.0	24
Sprayer/duster	1.0	63	1.0	44	1.0	57	1.0	61	1.0	52	1.0	57
Weeder	1.0	16	1.0	10	1.0	14	1.0	18	1.0	9	1.0	14
Seed drill/hoes	1.0	9	1.0	6	1.0	8	1.0	10	1.0	7	1.0	8
Infrastructure (Transport and Communication)												
Bicycle (non-agriculture)	1.0	13	1.0	13	1.0	13	1.0	12	1.0	13	1.0	13
Auto rickshaw	1.0	4	1.0	2	1.0	3	1.0	5	1.0	2	1.0	3
Two wheelers	1.1	79	1.0	72	1.0	77	1.1	78	1.0	75	1.0	77
Four wheelers	1.0	3	1.0	3	1.0	3	1.0	4	1.1	3	1.0	3
Non-farm assets (life-style goods)												
Sewing machine	1.0	6	1.0	4	1.0	5	1.0	6	1.0	5	1.0	5
TV	1.0	89	1.0	89	1.0	89	1.0	90	1.0	88	1.0	89
Computer/laptop	1.1	7	1.1	3	1.1	6	1.1	7	1.0	4	1.1	6
Refrigerator	1.0	44	1.0	39	1.0	43	1.0	43	1.0	42	1.0	43
LPG with Cooking set	1.0	93	1.0	95	1.0	93	1.0	93	1.0	94	1.0	93
Smartphone	1.2	81	1.1	69	1.1	77	1.1	81	1.1	72	1.1	77
Basic phone		100		100		100		100		100		100

Note: % indicates the proportion of households within each category.

TABLE 4.9
Asset Value and Poverty Status of Households

	Year 2020						Year 2022					
Total Assets (Value)	**Non-poor**		**Poor**		**Total**		**Non-poor**		**Poor**		**Total**	
Assets	**Mean**	%	**Mean**	%	**Mean**	%	**Mean**	%	**Mean**	%	**Mean**	%
Residential house value (Rs.)	6,51,503	99	5,07,607	99	6,03,743	99	6,22,644	99	5,82,375	99	6,03,743	99
Number of rooms (no)	3	99	3	99	3	99	3	99	3	99	3	99
Cultivated land value (Rs.)	66,43,468	95	42,41,476	96	58,40,446	96	67,84,451	95	47,78,255	96	58,40,446	96
Cattle shed/farmhouse value (Rs.)	53,736	18	45,857	13	51,641	17	52,332	18	50,695	15	51,641	17
Farm pond value (Rs.)	78,308	5	70,409	5	75,959	5	91,394	4	63,537	6	75,959	5
Storage house value (Rs.)	3,65,000	1	36,571	1	2,29,765	1	1,97,286	1	2,52,500	2	2,29,765	1
Total assets(value)	70,12,528	100	46,80,761	100	62,37,462	100	71,06,366	100	52,50,962	100	62,37,462	100

%, percent of households owning the asset.

TABLE 4.10
Ownership of Assets by Tenancy Status (2020–2021)

	Tenancy Status					
	Owner-Farmer		Pure Tenant-Farmer		Total	
Number of farmers	1,348		72		1,420	
	Mean	%	Mean	%	Mean	%
Land-Related Assets						
Cultivated land (acres)	5.0	100	4.7	17	5.0	96
Cattle shed/farmhouse (no.)	1.0	18	1.0	6	1.0	18
Farm pond (no)	1.0	6	1.0	4	1.0	6
Storage house (no)	1	1	1	2	1	1
Livestock						
Working cattle/buffalo (agriculture)	2	22	2	17	2	20
Milk cattle/buffalo	2	39	2	32	2	35
Young stock	1.6	15	1.5	14	1.6	15
Sheep	10.0	5	12.7	4	10.1	5
Goats	8.3	2	3.5	3	8.0	2
Poultry	8.0	9	3.8	6	7.9	9
Agricultural Assets (Equipment and Machinery)						
Tractor/trolley	1.0	16	1.0	4	1.0	15
Power tiller	1.0	5	1.0	1	1.0	4
Bullock cart	1.0	9	1.0	6	1.0	9
Wooden/iron plough	1.0	25	1.0	11	1.0	24
Sprayer/duster	1.0	58	1.1	44	1.0	57
Weeder	1.0	14	1.0	14	1.0	14
Seed drill/hoes	1.0	8	1.0	6	1.0	8
Infrastructure (Transportation and Communication)						
Bicycle (non-agriculture)	1.0	13	1.0	13	1.0	13
Auto rickshaw	1.0	3	1.0	4	1.0	3
Two wheelers	1.0	76	1.1	82	1.0	77
Four wheelers	1.0	4	1.0	1	1.0	3
Smartphone	1.1	77	1.1	76.4	1.1	77
Basic phone		100		100		100
Non-farm Assets (Lifestyle Goods)						
Sewing machine	1.0	5	1.0	6	1.0	5
TV	1.0	89	1.0	90	1.0	89
Computer/laptop	1.1	6	1.0	6	1.1	6
Refrigerator	1.0	43	1.0	36	1.0	43
LPG with cooking set	1.0	93	1.0	94	1.0	93

Note: % indicates the proportion of households within each category.

TABLE 4.11
Total Asset Value by Pure Tenant Status in Households

	Tenancy Status					
	Owner-Farmer		Tenant-Farmer		Total	
Assets	**Mean**	**%**	**Mean**	**%**	**Mean**	**%**
Residential house value (Rs.)	6,03,789	99	6,02,857	97	6,03,743	99
Number of rooms (no)	3	99	3	97	3	99
Cultivated land value (Rs.)	58,62,286	100	35,80,769	18	58,40,446	96
Cattle shed/farmhouse value (Rs.)	51,863	17	38,750	6	51,641	17
Farm pond value (Rs.)	76,315	5	50,000	1	75,959	5
Storage house value (Rs.)	2,29,765	1		0	2,29,765	1
Total assets value (Rs.)	65,02,330	100	12,78,542	100	62,37,462	100

Note: % indicates the proportion of households within each category.

Irrigation Is Essential

We spoke to both dryland and irrigated farmers and were astounded by the huge differences in their assets and wealth. It was clear that the availability of water was the key factor to this disparity. Sixty-two percent of farmers who had access to irrigation systems, on the contrary, flourished with larger cultivated lands, more cattle sheds, and farm ponds (Table 4.12). They also owned numerous farming tools and equipment. However, dryland farmers often maintained more cattle and livestock, which may compensate for their smaller cultivated land holdings. Dryland farmers generally invest less, as they expect lower returns on investment (Lauer and Sanderson, 2000; Tiwari et al., 2019; Kuchimanchi et al., 2021).

It was not just in terms of farming resources; the contrast was also apparent in wealth (Table 4.13). The value of assets owned by irrigated farmers amounted to Rs. 73 lakhs, a considerable contrast to the Rs. 45 lakhs possessed by dryland farmers. This was largely due to the consistent and high-yielding crops that irrigation provided, leading to better economic conditions.

COMPARISONS AMONG STATES

In the overall picture of the states, we noticed that the farmers in Andhra Pradesh had a diverse range of assets, especially in areas growing commercial crops like chilli, where they could afford high-input agriculture. Karnataka's farmers had bigger landholdings, so they had more agricultural assets because they used more machinery (Table 4.14). In Telangana, despite smaller landholdings, farmers were resourceful, especially in the chilli-growing regions, which brought higher profits but also higher risks. Rythu Bandhu scheme in Telangana helped farmers a lot by giving them money for farming, regardless of how much land they owned. This made their assets' worth increase more and improved their financial situation (Table 4.15).

TABLE 4.12
Ownership of Physical Assets Owned by Irrigated and Unirrigated Farmers (2021)

	Irrigation Status					
	Dry Land Farmers		Irrigated Farmers		Total	
Number of Farmers	544		876		1,420	
	Mean	%	Mean	%	Mean	%
Land-Related Assets						
Cultivated land (acres)	4.6	89	5.2	100	5.0	96
Cattle shed/farmhouse (no)	1.0	13	1.0	20	1.0	18
Farm pond (no)	1.0	3	1.0	8	1.0	6
Storage house (no)	1	1	1	2	1	1
Livestock						
Working cattle/buffalo (agriculture)	2	22	2	17	2	20
Milk cattle/buffalo	2	39	2	32	2	35
Young stock	1.4	13	1.6	16	1.6	15
Sheep	12.8	4	8.9	5	10.1	5
Goats	14.3	1	6.2	3	8.0	2
Poultry	4.2	7	9.5	10	7.9	9
Agricultural Assets (Equipments and Machinery)						
Tractor/trolley	1.0	14	1.0	16	1.0	15
Power tiller	1.0	4	1.1	4	1.0	4
Bullock cart	1.0	7	1.0	10	1.0	9
Wooden/iron plough	1.0	20	1.0	27	1.0	24
Sprayer/duster	1.0	49	1.0	62	1.0	57
Weeder	1.0	15	1.0	14	1.0	14
Seed drill/hoes	1.0	9	1.0	8	1.0	8
Infrastructure (Transportation and Communication)						
Bicycle (non-agriculture)	1.0	8	1.0	15	1.0	13
Auto rickshaw	1.0	2	1.0	4	1.0	3
Two wheelers	1.0	75	1.0	78	1.0	77
Four wheelers	1.0	2	1.0	4	1.0	3
Smartphone	1.1	73	1.2	79	1.1	77
Basic phone		100		100		100
Non-farm Assets (Lifestyle Goods)						
Sewing machine	1.0	5	1.0	6	1.0	5
TV	1.0	91	1.0	88	1.0	89
Computer/laptop	1.1	4	1.1	7	1.1	6
Refrigerator	1.0	41	1.0	44	1.0	43
LPG with cooking set	1.0	95	1.0	92	1.0	93

Note: % indicates the proportion of households within each category.

TABLE 4.13
Assets Value by Irrigation Status

	Dryland		Irrigated Land		Total	
Assets	**Mean**	**%**	**Mean**	**%**	**Mean**	**%**
Residential house value (Rs.)	5,14,578	98	6,58,307	99	6,03,743	99
Number of rooms (no)	3	98	3	99	3	99
Cultivated land value (Rs.)	44,28,495	89	66,24,863	100	58,40,446	96
Cattle shed/farmhouse value (Rs.)	43,639	13	55,133	19	51,641	17
Farm pond value (Rs.)	40,286	3	84,283	7	75,959	5
Storage house value (Rs.)	25,200	1	3,15,000	1	2,29,765	1
Total assets value (Rs)	45,41,676	100	72,90,553	100	62,37,462	100

Note: % indicates the proportion of households within each category.

TABLE 4.14
State-Wise Ownership of Assets

	State Name							
	Telangana		Karnataka		Andhra Pradesh		Total	
	Mean	**%**	**Mean**	**%**	**Mean**	**%**	**Mean**	**%**
Land-Related Assets								
Cultivable land (acres)	3.6	99	6.8	95	4.6	94	5.0	96
Cattle shed/farmhouse (no.)	1.0	25	1.1	15	1.0	14	1.0	18
Farm pond (no)	1.0	12	1.0	7	1.0	0	1.0	6
Storage house (no)	1.1	4	1.0	0	1.0	1	1.1	1
Livestock								
Working cattle/buffalo (agriculture)	1.9	26	2.1	27	2.0	7	2.0	20
Milk cattle/buffalo	1.6	37	2.1	37	2.1	32	1.9	35
Young stock	1.5	15	1.6	14	1.5	15	1.6	15
Sheep	9.8	4	10.9	9	4.0	1	10.1	5
Goats	2.9	3	5.0	2	23.3	1	8.0	2
Poultry and duck	8.5	15	11.8	5	3.8	7	7.9	9
Agricultural Assets (Farm Machinery and Equipment)								
Tractor/trolley	1.0	12	1.0	16	1.0	18	1.0	15
Power tiller	1.0	3	1.1	5	1.0	5	1.0	4
Bullock cart	1.0	10	1.0	13	1.0	4	1.0	9
Wooden/iron plough	1.0	29	1.0	27	1.0	18	1.0	24

(Continued)

TABLE 4.14 (*Continued*)
State-Wise Ownership of Assets

	State Name							
	Telangana		Karnataka		Andhra Pradesh		Total	
	Mean	%	Mean	%	Mean	%	Mean	%
Sprayer/duster	1.0	61	1.0	58	1.0	52	1.0	57
Weeder	1.0	8	1.0	25	1.0	9	1.0	14
Seed drill/hoes	1.1	2	1.0	16	1.0	7	1.0	8
Infrastructure (Transport and Communication)								
Bicycle (non-agriculture)	1.0	21	1.0	11	1.0	6	1.0	13
Auto rickshaw	1.0	3	1.0	3	1.0	3	1.0	3
Two wheelers	1.0	77	1.1	76	1.0	78	1.0	77
Four wheelers	1.1	4	1.0	5	1.0	1	1.0	3
Non-Farm Assets (Life Style)								
Sewing machine	1.0	7	1.0	7	1.0	3	1.0	5
TV	1.0	89	1.0	84	1.0	95	1.0	89
Computer/laptop	1.1	8	1.1	3	1.0	6	1.0	6
Refrigerator	1.0	52	1.0	32	1.0	44	1.0	43
LPG with cooking set	1.0	90	1.0	93	1.0	97	1.0	93
Smartphone	1.1	73	1.2	84	1.1	73	1.1	77
Basic phone	1.0	52	1.0	62	1.0	85	1.0	66

Note: % indicates the proportion of households within each category.

TABLE 4.15
Assets Value (in Rs.) Among States

	State Name							
	Telangana		Karnataka		Andhra Pradesh		Total	
	Mean	%	Mean	%	Mean	%	Mean	%
Residential house value	7,13,043	100	5,52,406	100	5,48,491	97	6,03,743	97
Cultivable land value	50,63,998	99	77,84,912	95	46,55,089	94	58,40,446	96
Cattle shed/ farmhouse value	54,828	22	53,379	15	44,769	14	51,641	17
Farm pond value	70,810	9	82,719	7			75,959	5
Storage house value	2,29,000	3	6,00,000	0	50,000	0	2,29,765	1
Total assets value	57,36,582	100	79,46,403	100	50,08,531	100	62,37,462	100

Note: % indicates the proportion of households within each category.

This experience opened our eyes to the challenges and inequalities within the world of farming, and we felt a sense of admiration for these hardworking farmers. One of the main reasons for these inequalities was land. We strongly felt that these trends are harmful and must be recognized and addressed (Anand and Thampi, 2016). It also left us with important insights for policymakers to consider to support and uplift rural communities.

REFERENCES

Agriculture Census Division. (2018). https://agcensus.da.gov.in/acindia.html

Anand I, Thampi A. (2016). Recent Trends in Wealth Inequality in India. *Economic and Political Weekly*, 51(50), pp. 59–67.

Aryal JP, Maharjan S, Erenstein O, Rahut DB. (2019). Understanding Factors Associated with Agricultural Mechanization: A Bangladesh Case. *World Development Perspectives*, 13, pp. 1–9.

Aryal JP, Thapa G, Simtowe F, Rahut DB. (2021). Mechanisation of Small-Scale Farms in South Asia: Empirical Evidence Derived from farm Households Survey. *Technology in Society*, 65, p. 101591.

Bansal G. (2020). Tenancy and Accumulation: A Study of the Capitalist Farm Sector in Punjab. *Review of Agrarian Studies*, 10(2), 28–47.

Baruah S, Mohanty S, Rola AC. (2022). Empowering Women Farmers through Collective Action: A Case Study of Khanizpur Hamlet, Odisha. *Gender, Technology and Development*, 26(1), pp. 58–73.

Bhattacharya S. (2019). Agricultural Tenancy in Contemporary Punjab: A Study Based on National Sample Survey Data. *Review of Agrarian Studies*, 9(2), pp. 80–91.

Brown AG. (ed.) (2003) The Livestock Revolution: A Pathway from Poverty? In *Proceedings of a Conference Held at the ATSE Crawford Fund, Parliament House, Canberra*. ATSE Crawford Fund, Canberra.

Catley A. (2008). The Growing Demand for Livestock. Will Policy and Institutional Changes Benefit Poor People? *ID21 Insights*, 72, pp. 1–2.

Delgado C. (2003). Rising Consumption of Meat and Milk in Developing Countries Has Created a New Food Revolution. *Journal of Nutrition*, 133, pp. 3907S–3910S.

Desai R. (2016). A budget amid agrarian crisis. Research Unit for Political Economy. https://rupeindia.wordpress.com/2016/04/13/a-budget-amid-agrarian-crisis-pt-iii/

Deshpande RS. (2002). Suicide by Farmers in Karnataka: Agrarian Distress and Possible Alleviatory Steps. *Economic and Political Weekly*, 37, pp. 2601–2610.

Hossain M, Malek MA, Hossain MA., Reza MH., Ahmed MS. (2019). Agricultural Microcredit for Tenant Farmers: Evidence from a Field Experiment in Bangladesh. *American Journal of Agricultural Economics*, 101(3), pp. 692–709.

ILRI. (2007). *ILRI Annual Report 2007 – Markets that Work*. International Livestock Research Institute (ILRI), Nairobi Kenya, ILRI.

ILRI. (2011). *Livestock: A Pathway Out of Poverty. Interim Strategy of the International Livestock Research Institute, 2011–2012*. Nairobi, Kenya, ILRI.

Jansen HG, Pender J, Damon A, Wielemaker W, Schipper R. (2006). Policies for Sustainable Development in the Hillside Areas of Honduras: A Quantitative Livelihoods Approach. *Agricultural Economics*, 34(2), pp. 141–153.

Kadaraiah A, Basavaprabhu J, Shivananda P, Yarazari HM, Nandini, CPN. (2022). Perception of Farmers towards Custom Hiring Service Centres in Tumakuru District of Karnataka. *Indian Journal of Extension Education*, 58(2), pp. 209–212.

Kuang F, Jin J, He R, Ning J, Wan X. (2020). Farmers' Livelihood Risks, Livelihood Assets and Adaptation Strategies in Rugao City, China. *Journal of Environmental Management*, 264, p. 110463.

Kuchimanchi BR, Van Paassen A, Oosting SJ. (2021). Understanding the Vulnerability, Farming Strategies and Development Pathways of Smallholder Farming Systems in Telangana, India. *Climate Risk Management*, 31, p. 100275.

Lauer S, Sanderson M. (2020). Irrigated Agriculture and Human Development: A County-Level Analysis 1980–2010. *Environment, Development and Sustainability*, 22, pp. 4407–4423.

Lindenberg M. (2002). Measuring Household Livelihood Security at the Family and Community Level in the Developing World. *World Development*, 30(2), pp. 301–318.

Mehta CR, Chandel NS, Jena PC, Jha A. (2019). Indian Agriculture Counting on Farm Mechanization. *Agricultural Mechanization in Asia, Africa and Latin America*, 50(1), pp. 84–89.

Mohanty BB. (2005). We Are Like the Living Dead: Farmer Suicides in Maharashtra, Western India. *Journal of Peasant Studies*, 32(2), pp. 243–276.

Mohanty BB, Lenka PK. (2019). Small Farmers' Suicide in Odisha. *Economic and Political Weekly*, IV(22), pp. 51–59.

Patnaik, U. (2024). The Many Republics of Hunger. Agrarian South: Journal of Political Economy, 13(4), pp. 489–505.

Pica-Ciamarra U. (2009). Livestock-Poverty Linkages in Latin America. *Livestock Research for Rural Development*, 21(1), Article number 11.

Rao S. (2017). Caste Discrimination and Agricultural Performance in India. *Economic and Political Weekly*, 52(25 & 26), pp. 32–38.

Reddy AA. (2021). Assuring Farmers Income in the Context of New Farm Laws: Issues and the Way Forward. In *2021 Conference*, August 17–31, 2021, Virtual (No. 315243). International Association of Agricultural Economists.

Reddy DN, Reddy Amarender A, Nagaraj N, Bantilan C. (2014). *Rural Non-Farm Employment and Rural Transformation in India*. Working Paper Series No. 57. International Crops Research Institute for the Semi-Arid Tropics, Telangana, India, 28 pp.

Sarkar A. (2020). Agricultural Mechanization in India: A Study on the Ownership and Investment in Farm Machinery by Cultivator Households across Agro-Ecological Regions. *Millennial Asia*, 11(2), pp. 160–186.

Sekaran U, Lai L, Ussiri DA, Kumar S, Clay S. (2021). Role of Integrated Crop-Livestock Systems in Improving Agriculture Production and Addressing Food Security–A Review. *Journal of Agriculture and Food Research*, 5, 100190.

Sedai AK, Vasudevan R, Pena AA. (2021). Friends and Benefits? Endogenous Rotating Savings and Credit Associations as Alternative for Women's Empowerment in India. *World Development*, 145, p. 105515.

Shah E. (2012). A LifeWasted Making Dust: Affective Histories of Dearth, Death, Debt and Farmers' Suicides in India. *The Journal of Peasant Studies*, 39(5), pp. 1159–1179.

Sidhu RS, Singh S, Bhullar AS. (2011). Farmers' Suicides in Punjab: A Census Survey of the Two Most Affected Districts. *Economic and Political Weekly*, 46(26–27), pp. 131–137.

Stringer LC, Fraser EDG, Harris D, Lyon C, Pereira L, Ward CFM, Simelton, E. (2020). Adaptation and Development Pathways for Different Types of Farmers. *Environmental Science & Policy*, 104, pp. 174–189.

Tiwari PS, Singh KK, Sahni RK, Kumar V. (2019). Farm Mechanization–Trends and Policy for its Promotion in India. *Indian Journal of Agricultural Sciences*, 89(10), pp. 1555–1562.

Udoh EJ, Akpan SB, Uko EF. (2017). Assessment of Sustainable Livelihood Assets of Farming Households in Akwa Ibom State, Nigeria. *Journal of Sustainable Development*, 10(4), pp. 83–96.

Vasavi AR. (1999). Agrarian Distress in Bidar: Market, State and Suicides. *Economic and Political Weekly*, 34(32), pp. 2263–2268.

Vu HT, Tran D, Goto D, Kawata K. (2020). Does Experience Sharing Affect Farmers' Pro-Environmental Behavior? A Randomized Controlled Trial in Vietnam. *World Development*, 136, p. 105062.

Yang L, Liu M, Lun F, Min Q, Zhang C, Li H. (2018). Livelihood Assets and Strategies among Rural Households: Comparative Analysis of Rice and Dryland Terrace Systems in China. *Sustainability*, 10, p. 2525.

5 Story of Farmer's Land

We had always known that land was not only a precious asset, but its ownership was a symbol of prestige and security for farmers. A farmer told us "*To most people, land is just a place to work. But it's much more than that for the farmers like us. The land is our **L**ife, **A**mbition, **N**ative and **D**etermination. So, if I lose my land, I lose everything.*" However, the scarcity of land poses a formidable challenge for small farmers. Other side, many farmers told us they could not work as daily wagers due to a social "stigma." So, hiring land on an annual lease or leased in land was one of the few options for them. As we discussed their land, their land's story and the unique system of land ownership and sharing that existed in their communities made us curious.

As we delved deeper into this practice and examined the overall data on land ownership across India, we were struck by the diverse narratives that emerged from each state. From the fertile plains of Punjab to the rugged hills of Himachal Pradesh and the expansive paddy fields of West Bengal, every state had its own distinct story (Bardhan, 1984; Binswanger et al., 1993).

In states like Himachal Pradesh, where land was plentiful, farmers enjoyed sufficient harvests, and land-sharing practices were minimal. Conversely, in states like Bihar, where land was scarce, leasing became the primary means of production and a vital source of livelihood for farmers (Agriculture-Census-Division, 2019; Goyal et al., 2022). Large-scale farmers in Punjab and Haryana often leased their land to smaller farmers in need. In contrast, in states like West Bengal and Kerala, where smallholder farmers predominated, it was common for them to share their land to supplement their income (Chakravorty, 2013). Despite the varied experiences of these farmers, they all shared a common goal: to provide for their families (Agarwal, 1994).

RELIGION MATTERS

Farmers' information and our calculation revealed that 95% of farmers owned their land as a symbol of their hard work and dedication (Table 5.1). However, it surprised us that only a small percentage of farmers leased out their land. It was interesting to see the dynamic between ownership and leasing among different religious groups. Christians had the highest percentage of farmers leased in land, while Hindus had the lowest. Our analysis also showed that, on average, farmers owned 5 acres of land. But, due to the higher proportion of leased in land and lower proportion of leased-out land, the cultivated area increased to 6.9 acres. It was heartwarming to see how the farmers shared their knowledge, land, and resources. They all agreed that leasing land was not just about monetary gain but also about cultivating relationships and helping each other.

DOI: 10.1201/9781003582878-5

TABLE 5.1
Land Tenancy Status Based on Religion (Area in acres)

	Religion							
	Hindu		Muslim		Christian		Total	
	Mean	%	Mean	%	Mean	%	Mean	%
Tenancy Status								
Own land 2021	5.0	95	5.0	90	2.4	87	5.0	95
Own land 2022	5.0	95	5.0	90	2.4	87	5.0	95
Leased in 2021	7.6	28	12.4	33	4.9	39	7.7	28
Leased in 2022	7.5	29	13.1	33	4.3	43	7.6	29
Leased out 2021	9.4	1	30.0	2	1.0	4	10.1	1
Leased out 2022	6.1	1	16.0	4	1.0	4	7.3	1
Cultivated area 2021	6.9	100	7.7	100	4.0	100	6.9	100
Cultivated area 2022	7.0	100	7.8	100	3.9	100	6.9	100

Note: % indicates the proportion of households within each category.

SOCIAL STRATA

In rural India, agricultural landholding and food security are closely connected with social hierarchy or caste (Goli et al., 2021). The association between landholding and caste presents some important irregularities and creates a large variation (Fischer, 1997). Though we had heard of this practice before, we truly understood its impact until we visited these villages. The village was divided into different sections, each occupied by families belonging to a specific caste. As we spoke to the farmers in the village, we realized the deep-rooted impact of this system where this division extended to land ownership as well. In these villages, upper castes, mostly forward caste (FC), dominate the farming landscape, leading to higher incomes and better living standards. On the contrary, the scheduled tribe (ST) and scheduled caste (SC) farmers struggled to make ends meet, with limited land and resources at their disposal (Table 5.2).

The average acreage of cultivated land by FC farmers was 5.7 acres, which was almost double that of SC at 2.9 acres.

This disparity in land ownership was a direct result of the caste-based system that had been in place for centuries. But what was even more concerning was the presence of social prejudices and discrimination in the land lease market. The data revealed that FC and other backward class farmers were more likely to lease in land, while ST and SC farmers were more likely to lease out land.

"*Why are you not raising your voice?*" One of our researchers asked a lower-caste farmer. He hesitantly stated, "*We are not vocal enough to challenge this arrangement.*"

This entire situation exacerbated the existing inequalities and widened the gap between different castes even further.

TABLE 5.2
Land Tenancy Status by Social Group (Area in acres)

	Caste									
	Scheduled Castes		Scheduled Tribes		Other Backward Classes		Forward Caste		Total	
	Mean	%	Mean	%	Mean	%	Mean	%	Mean	%
Tenancy Status										
Own land 2021	2.9	90	3.9	97	5.0	94	5.7	96	5.0	95
Own land 2022	2.9	90	3.9	97	4.9	94	5.7	96	5.0	95
Leased in 2021	5.8	31	4.6	16	7.5	29	8.7	30	7.7	28
Leased in 2022	5.5	32	4.7	16	7.5	30	8.6	31	7.6	29
Leased out 2021	2.0	2	12.0	1	11.5	1	11.6	2	10.1	1
Leased out 2022	2.0	2		0	12.0	1	6.8	1	7.3	1
Cultivated area 2021	4.4	100	4.6	100	6.8	100	8.1	100	6.9	100
Cultivated area 2022	4.4	100	4.7	100	6.9	100	8.1	100	6.9	100

Note: % indicates the proportion of households within each category.

LAND CONFLICT

When we focused on the farmers' size and its impact on agricultural land ownership, something caught our attention. Many marginal farmers and a significant portion of small farmers leased in land. This meant that they did not own enough cultivated land; instead, they rented it from other landowners. The average leased in land size ranged from 6.3 to 6.9 acres. We delved deeper into this phenomenon and discovered that many small farmers were forced to lease land to sustain their livelihoods. It was a vicious cycle for these farmers. They lacked the resources to purchase land, so they had to lease it. However, this meant that they faced challenges in negotiating fair lease agreements, accessing credit and resources, and dealing with uncertain land tenure arrangements. Further, variations in agricultural land tenancy laws among states either prohibit leasing or permit it under specific conditions, making leasing difficult for small and marginal farmers to engage in leasing agreements (Singhi & Arora, 2008). It was because none of the states in the country (except a few) were well equipped with laws to guide the terms of the contract (Planning Commission, 2006; FAO, 2024).

We also noticed that 14% of large landowners and about one-fifth of medium and semi-medium landowners also leased-in land (Table 5.3). The difference was that they leased in significantly larger land sizes ranging from 7.6 to 13.8 acres, which allowed them to expand their operations without the need to purchase additional land. It signified that owning larger operational holding sizes allowed farmers to benefit from economies of scale. On the contrary, those who leased land struggled to make ends meet.

TABLE 5.3
Tenancy Status by Land Size Category (Acre)

	Marginal		Small		Semi-Medium		Medium		Large		Total	
	Mean	%	Mean	%	Mean	%	Mean	%	Mean	%	Mean	%
Tenancy Status												
Own land 2021	1.1	66	1.8	100	3.3	99	6.4	100	19.6	100	5.0	95
Own land 2022	0.9	64	1.8	100	3.3	100	6.4	100	19.6	100	5.0	95
Leased in 2021	6.8	59	6.4	30	8.0	24	8.7	22	13.8	14	7.7	28
Leased in 2022	6.9	59	6.3	30	7.6	25	8.9	22	13.8	14	7.6	29
Leased out 2021	5.0	1	1.8	1	24.7	1	5.7	2	17.3	3	10.1	1
Leased out 2022		0	3.5	1		0	4.4	2	16.7	3	7.3	1
Cultivated area 2021	4.7	100	3.8	100	5.2	100	8.3	100	21.1	100	6.9	100
Cultivated area 2022	4.8	100	3.8	100	5.2	100	8.4	100	21.0	100	6.9	100

Note: % indicates the proportion of households within each category.

POVERTY STATUS MATTERS

We spent a lot of time interviewing farmers from all walks including poor and non-poor to explore how their economic status influenced their involvement in land leasing. The discussion with the farmers showed a clear difference in land ownership and cultivated areas between these two groups. On average, poor farmers own less land and less leased-in land, resulting in a smaller cultivated area compared to non-poor farmers (Table 5.4). This further limits their ability to increase their productivity and profitability.

The most significant barrier for poor farmers was their limited capacity to take on additional leased in land. With only 3.8 acres of land on average, they are unable to benefit from economies of scale and other advantages that come with leasing more land. Non-poor farmers, on the contrary, can easily take on more leased in land, allowing them to increase their productivity and profitability.

TENANCY IMPACT

As we immersed ourselves in gaining a deeper understanding of landless tenant farmers' struggles and aspirations, we discovered that they rely on leasing small plots of land from landowners to cultivate crops or raise livestock (Table 5.5). They

TABLE 5.4
Tenancy Status by Poverty Level (Area in acres)

	2020–2021						2021–2022					
	Non-Poor		Poor		Total		Non-Poor		Poor		Total	
	Mean	%	Mean	%	Mean	%	Mean	%	Mean	%	Mean	%
Tenancy Status												
Own land 2021	5.5	95	3.8	95	5.0	95	5.5	95	4.3	95	5.0	95
Own land 2022	5.5	95	3.8	96	5.0	95	5.5	95	4.3	95	5.0	95
Leased in 2021	8.4	29	6.2	28	7.7	28	8.7	28	6.6	29	7.7	28
Leased in 2022	8.4	29	6.2	29	7.6	29	8.6	28	6.6	30	7.6	29
Leased out 2021	11.8	1	5.2	1	10.1	1	13.5	1	5.4	1	10.1	1
Leased out 2022	7.9	1	1.0	0	7.3	1	7.7	1	6.3	0	7.3	1
Cultivated area 2021	7.7	100	5.4	100	6.9	100	7.7	100	6.0	100	6.9	100
Cultivated area 2022	7.7	100	5.5	100	6.9	100	7.7	100	6.1	100	6.9	100.0

Note: % indicates the proportion of households within each category.

TABLE 5.5
Owner-Tenant and Landless Tenant Status 2022 (Acre)

	Tenancy status 2022					
	Owner-cum-tenant		Landless Tenant		Total	
Tenancy Status 2022	Mean	%	Mean	%	Mean	%
Own land 2021	5.0	100		0	5.0	95
Own land 2022	5.0	100		0	5.0	95
Leased in 2021	7.7	25	7.6	90	7.7	28
Leased in 2022	7.7	26	7.5	92	7.6	29
Leased out 2021	10.4	1	5.0	1	10.1	1
Leased out 2022	7.3	1		0	7.3	1
Cultivated area 2021	6.9	100	7.3	100	6.9	100
Cultivated area 2022	6.9	100	7.5	100	6.9	100

Note: % indicates the proportion of households within each category.

said that the constant struggle between land ownership and the tenancy status was their biggest issue.

"*We dream of owning our land, but we cannot afford it. We must rely on leasing land, which does not provide any security as we have no legal rights to it*," said one farmer.

This tenant farming results in declining land productivity as well as increasing land degradation (Mythili & Goedecke, 2016). The tenant farmers were reluctant to invest adequately as they did not have secure future rights to the land (FAO, 1994). However, in recent years, some relatively affluent tenant farmers started leasing larger parcels of land to cultivate commercial crops such as chillies, cotton, and paddy. This has created a divide within the community, with some farmers becoming more prosperous while others continue to struggle. During our discussion, we could see the reflection of the complexities and challenges of the land owners and tenant farmers as it is influenced by factors such as caste, class, and gender. Many landless tenant farmers encounter significant challenges, including limited access to credit, insecure tenure, and low productivity. Though we saw some evidence about government policies, land reforms, tenancy laws, and agricultural development programmes and programmes supporting these farmers, there is still a long way to go in addressing the insecurity faced by these farmers.

IRRIGATION IS ESSENTIAL

We met both a group of dryland farmers and irrigated farmers, and we got a chance to study the differences in land tenancy arrangements between them. The dryland farmers explained to us how their land tenancy arrangements were influenced by the unpredictable climate conditions in their area. Due to the reliance on rainfall for crop production, the lease periods were shorter, usually for a single growing season. This was to minimize the risks associated with unpredictable weather patterns, which could greatly impact their productivity. Most of them (Kareemulla et al., 2021) have the capital to invest in irrigation infrastructure. Further, the ignorance of the farmers due to the inability to maintain makes the land more and more vulnerable to degradation (Vijayan, 2016).

On the other side, where farmers had access to a dependable water source for irrigation, the land tenancy arrangements were completely different. The productivity of the land was more consistent, leading to longer lease periods and more stable arrangements. However, irrigated land was more expensive to lease, which put challenges for farmers who were unable to afford the high rental rates (Deshpande, 2022). But what struck us the most was the impact of these arrangements on the productivity and profitability of the farmers. The average cultivated land among irrigated farmers was slightly higher, and they also possessed more owned land compared to dryland farmers (Table 5.6). This raised concerns about the productivity and profitability of the latter, as they faced greater uncertainty and lower yields due to rainfall variability. As we continued our research, we realized that it was a complex system where the differences in land tenancy arrangements were not just a result of geographical factors but also influenced by cultural and social norms, inheritance practices, and even demographic factors such as population dynamics and urbanization.

TABLE 5.6
Land Tenancy Status for Dryland and Irrigated Farmers (Acreage)

Irrigation Status, 2022	Classification Based on Irrigation 2022					
	Dry Land		Irrigated Land		Total	
	Mean	%	Mean	%	Mean	%
Own land 2021	4.5	87	5.2	100	5.0	95
Own land 2022	4.5	87	5.2	100	5.0	95
Leased in 2021	6.8	38	8.6	23	7.7	28
Leased in 2022	6.8	39	8.6	23	7.6	29
Leased out 2021	5.9	2	15.9	1	10.1	1
Leased out 2022	6.8	1	7.6	1	7.3	1
Cultivated area 2021	6.5	100	7.1	100	6.9	100
Cultivated area 2022	6.6	100	7.1	100	6.9	100

Note: % indicates the proportion of households within each category.

TABLE 5.7
Tenancy Status of Farmers by State

	State Name							
	Telangana		Karnataka		Andhra Pradesh		Total	
	Mean	%	Mean	%	Mean	%	Mean	%
Total own land 2021	3.6	98	6.8	94	4.5	92	5.0	95
Total own land 2022	3.6	98	6.8	94	4.5	93	5.0	95
Total leased in 2021	6.5	21	10.3	21	7.0	43	7.7	28
Total leased in 2022	6.5	20	10.4	22	6.8	45	7.6	29
Total leased out 2021	1.7	1	16.5	2	3.6	1	10.1	1
Total leased out 2022	3.8	1	11.0	1	7.0	0	7.3	1
Total cultivated area 2021	4.9	100	8.6	100	7.2	100	6.9	100
Total cultivated area 2022	4.9	100	8.6	100	7.2	100	6.9	100

Note: % indicates the proportion of households within each category.

COMPARISONS AMONG STATES

We were amazed by the vast lands owned by farmers in Karnataka – an average of 6.8 acres of land, the highest among the three states (Table 5.7). What surprised us even more was that farmers were open to leasing even more land, averaging 10 acres each. On the contrary, farmers in Andhra Pradesh owned comparatively less land, they leased a high percentage of land, ranging from 43% to 45% because of government schemes that made it easier for them to lease land and boost their productivity. Despite the lowest land ownership in Telangana, farmers still managed to cultivate an average of 4.9 acres by leasing additional land, which might be because of the state's economic conditions and other opportunities available to them.

Overall, our research highlighted the struggles and differences faced by Indian farmers when it came to land ownership and renting. It sheds light on the complex web of factors that influence this system, and the challenges that farmers continue to face in their daily lives. We were hopeful that these findings would bring about positive change and lead to a more equitable distribution of land among farmers.

REFERENCES

Agarwal B. (1994). *A Field of One's Own: Gender and Land Rights in South Asia*. No. 58. Cambridge University Press.

Bardhan P. (1984). *Land, Labor, and Rural Poverty: Essays in Development Economics*. Columbia University Press.

Binswanger HP, Deininger K, Feder G. (1993). Power, Distortions, Revolt and Reform in Agricultural Land Relations. *Handbook of Development Economics*, 3, pp. 2659–2772.

Chakravorty S. (2013). *The Price of Land: Acquisition, Conflict, Consequence*. Oxford University Press.

Deshpande RS. (2022). Disaster Management in India: Are We Fully Equipped?. *Journal of Social and Economic Development*, 24(Suppl 1), pp. 242–281.

FAO. (1994). Land degradation in South Asia: Its severity causes and effects upon the people. Food and Agriculture Organisation. https://www.fao.org

FAO. (2024). FAO in India: India at a Glance. https://www.fao.org/india/fao-in-india/india-at-a-glance/en/

Fischer RJ. (1997). If Rain Doesn't Come: An Anthropological Study of Drought and Human Ecology in Western Rajasthan. *Sydney Studies in Society and Culture*, 14, Manohar Publishers, New Delhi, pp. 234.

Goli S, Rammohan A, Reddy SP. (2021). The Interaction of Household Agricultural Landholding and Caste on Food Security in Rural Uttar Pradesh, India. *Food Security*, 13, pp. 219–237.

Goyal Y, Choudhury PR, Ghosh RK. (2022). Informal Land Leasing in Rural India Persists Because It Is Credible. *Land Use Policy*, 120, 106299.

Kareemulla K, Krishnan P, Ravichandran S, Kumar B, Sharma S, Bhatta R. (2021). Spatiotemporal Analysis of Size and Equity in Ownership Dynamics of Agricultural Landholdings in India Vis-à-Vis the World. *Sustainability*, 13(18), p. 10225.

Mythili G, Goedecke J. (2016). Economics of Land Degradation in India. In: Nkonya, E., Mirzabaev A, von Braun J. (Eds.), *Economics of Land Degradation and Improvement – A Global Assessment for Sustainable Development*. Springer, Cham. https://doi.org/10.1007/978-3-319-19168-3_15.

Planning Commission. (2006). Report of the Working Group on Land Relations for Formulation of 11th Five Year Plan. (2006). Planning Commission, Govt. of India, New Delhi.

Singhi LC, Arora S. (2008). Land Tenancty and Contract Farming. In R.S. Deshpande (Ed.), *Contract Farming and Tenancy Reforms – Entangled without Tether* (pp. 79–86). Lal Bahadur Shastri National Academy of Administration, Mussoorie, New Delhi.

Vijayan R. (2016). Dryland Agriculture in India – Problems and Solutions. *Asian Journal of Environmental Science*, 11(2), pp. 171–177.

6 Exploring Farmers' Income

A Tale of Harvesting Hope

After learning about the farmers' assets and wealth, we delved into their income sources to uncover more about their livelihoods. We visited different farms and spoke to farmers from different backgrounds. We saw right away the struggles they faced and the determination with which they worked to provide for their families.

We met Shibani, an about 40-year-old woman farmer, who lives in a remote village in Telangana with her four children. Her husband works half the year in another state. She used to work as a labourer on the agricultural land and in the off-season in the Mahatma Gandhi National Rural Employment Guarantee Act (MGNREGA). She often walked for 5 km to work, and her children were home alone for the whole day. Her left hand was partially paralysed, but she worked as much as she could. She earned about Rs. 150 per day, much less than the standard wage due to her disability. During harvest season, she gathers crops and is paid in kind, keeping one-ninth of the produce. But this work lasts only about four weeks per year, and, overall, Shibani works less than two months annually with no guarantee of daily employment. But she and her husband were very hopeful for their children. However, she regrets not being able to educate them due to the poverty she strives to work as much as she can.

We were struck by the resilience and hard work of those farmers we met. They were the backbone of our country, yet they were deeply affected by the vicious cycle of poverty (Olawepo, 2010). Despite their crucial contribution, their income did not reflect the importance of their role. The farm income was very meagre and insufficient to pay even for their essentials (Sen and Bhatia, 2004). We discovered that the income disparity between large farmers compared to small and marginal farmers was substantial. Some farmers were earning substantially more than the average, while others were struggling to make ends meet because those who cultivate larger farm sizes tend to earn more income from their investments. According to the National Statistical Office, the average monthly income of a farm household in India was estimated to be around Rs. 6,426 in 2018, which is equivalent to USD 87 (approx.). But this figure varied a lot in different regions and farming groups because of differences in how crops are grown, access to water for irrigation, use of new technologies, government policies for buying crops, market setups, and other factors (Narayanamoorthy, 2017). As soon as we realized that farmers' income was contingent on various factors, we were determined to understand the factors that influenced their earnings and how they could be improved. The farmers told us that the quality and productivity of their land, the type and yield of crops they cultivated, the prices they received for their produce, and the availability and cost of agricultural inputs

DOI: 10.1201/9781003582878-6

were just some of the factors that influenced their earnings. Other factors such as weather conditions, government policies, and socio-economic aspects also played a significant role. Also, we found that the limited access to resources, technology, and markets was a significant factor contributing to the income disparity. However, we witnessed how technology and innovation, such as the use of improved seeds and fertilizers, could increase productivity and income.

EXPLORING DIVERSE STREAMS OF FARMER'S INCOME ACROSS TIME

Crop cultivation has long been the primary source of income for most farming families. This deep-rooted connection to the land typically generated an annual income of around Rs. 2 lakh, providing sustenance through changing seasons (Hazell and Rahman, 2014; Jayne et al., 2010). However, the traditional reliance on agriculture also posed significant challenges, as the success of each harvest was contingent on various unpredictable factors such as weather, pests, and market prices. The inherent volatility in farming has led many farmers to diversify into non-farm enterprises as a means of reducing risk and boosting income generation (Ellis, 2000; Barrett et al., 2001). This trend of diversification reflects a broader strategy adopted by rural households to secure their livelihoods amidst the uncertainties of agriculture (Reardon et al., 2001; Odoh et al., 2020; Nwibo et al., 2017).

As the farmers said, the MGNREGA was a significant source of income for them. This government scheme provided them with unskilled manual labour opportunities for at least 100 days a year and brought a steady income. It served as a lifeline for many of them. We were surprised to find that many farmers were also involved in self-employment, small businesses, and other agricultural activities such as animal husbandry and sericulture. Some even ventured into fruit and vegetable production, finding new opportunities for diversification. As we spoke to different farmers, we realized that they were resilient and adaptable, constantly seeking new opportunities to improve their income. Some relied on remittances and pensions for financial support, while others took up casual labour in construction activities or domestic work.

However, with the changing seasons, the story of rural households was also evolving. The wind of fortune could be very changeable and bring unpredictable shocks in the next year. In the year 2021–2022, all three states saw a dramatic decline in crop yields because of the failure of the chilli crop which was the main source of income for many farmers. The failure of this crop rippled through the entire rural economy.

DIGGING DEEPER INTO THE GROUPS

Religious Groups

It was an eye-opening experience for us as we delved deeper into the lives of these hard-working individuals. Upon closer inspection, we noticed the differences in the sources of income among the religious groups. While the Hindu and Muslim farmers heavily relied on agriculture, the Christian farmers had diversified their income sources to non-farm sectors especially MGNREGA (Tables 6.1 and 6.2). We were

TABLE 6.1
Income (Rs/Annum) by Religion (2020–2021)

Income Source	Religion						Total	
	Hindu		Muslim		Christian			
	Mean	%	Mean	%	Mean	%	Mean	%
Field crops	1,93,442	72	1,83,067	71	44,525	48	1,91,456	72
MGNREGA	10,869	65	9,183	69	8,429	91	10,749	66
Casual labourer in agriculture	21,432	48	22,200	56	13,786	61	21,310	51
Livestock	25,928	35	24,727	21	33,200	22	25,976	34
Business/self-employment	2,29,977	23	2,15,750	23	55,000	9	2,28,350	23
Casual labourer in construction/ non-agriculture	31,996	10	21,654	25	12,400	22	30,525	11
Remittances and pensions	27,453	9	22,800	8	30,000	39	27,485	9
Worked as domestic workers	49,962	2	45,333	12		0	49,094	2
Fruits and vegetables	1,80,000	1	3,80,000	2		0	2,13,333	0.4
Sericulture	50,000	0.4		0		0	50,000	0.4
Total household income	2,25,048	100	2,23,722	100	63,817	100	2,22,388	100

Note: % indicates the proportion of households within each category.

TABLE 6.2
Household Income (Rs./Annum) by Religion (2021–2022)

Income Source	Religion						Total	
	Hindu		Muslim		Christian			
	Mean	%	Mean	%	Mean	%	Mean	%
MGNREGA	11,323	65	6,683	69	8,048	91	11,069	65
Field crops	1,19,871	52	1,54,877	60	16,593	13	1,20,926	52
Casual labourer in agriculture	21,935	48	22,510	56	14,286	61	21,804	49
Livestock	26,296	34	27,545	21	37,750	17	26,420	34
Business/self-employment	2,38,212	23	2,17,417	23	55,000	9	2,36,293	23
Casual labourer in construction/ non-agriculture	30,270	10	22,108	25	12,400	22	29,025	11
Remittances and pensions	27,453	9	22,800	8	30,000	39	27,485	9
Worked as domestic workers	48,200	2	46,167	12		0	47,806	2
Fruits and vegetables	2,80,857	1	2,40,000	4		0	2,71,778	1
Sericulture	2,00,000	0.4		0		0	2,00,000	0.2
Total household income	1,51,879	100	1,87,350	100	43,990	100	1,51,431	100

Note: % indicates the proportion of households within each category.

surprised to see that a huge share of Christian farmers engaged as casual labour both in farm and non-farm sectors and relied on remittances and pensions for their livelihood earning a good amount of money.

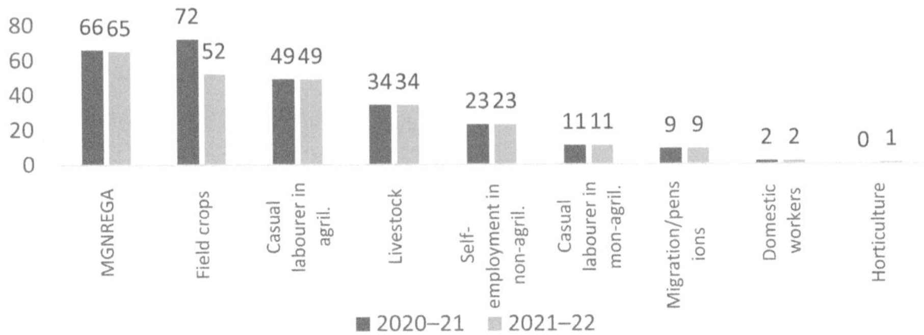

FIGURE 6.1 Percentage of farmers engaged in different economic activities.

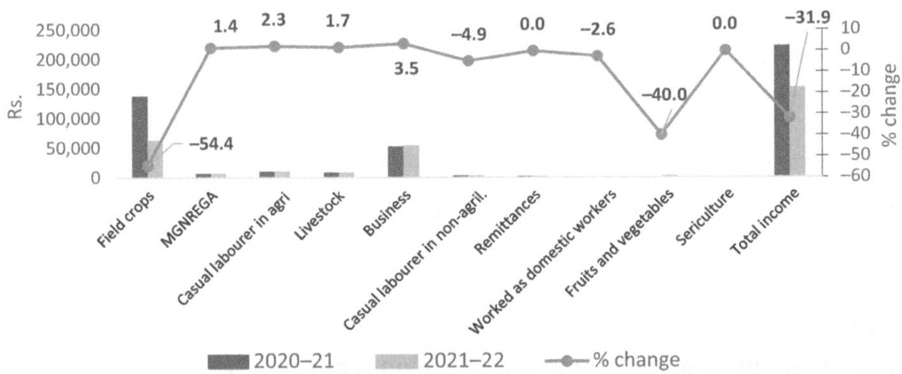

FIGURE 6.2 Household income (2020–2021 vs 2021–2022).

But our research took a mournful turn when we compared the income patterns of the previous year, 2020–2021, to the year 2021–2022 (Figure 6.1). There was a severe impact of crop loss. Field crop cultivation saw a 54% reduction, casual labours faced a 5% decrease, and even working as domestic workers saw a 3% decline (Figure 6.2). The only silver lining was a 3% increase in self-employment in non-agricultural activities.

After conducting numerous interviews, we were able to draw some interesting conclusions. We found that the dependency on agriculture and its income reduced drastically in 2021–2022 among all the castes (Tables 6.3 and 6.4). This was due to a huge exposure to crop loss in the later year. The general and OBC farmers were impacted mostly. But the average income of farmers from livestock was the same in both years. However, the income from self-employment in non-agriculture had increased in 2021–2022. One interesting thing was even during the disaster in 2021–2022, many ST farmers were able to earn income from field crops, while others were not able to do so (Figure 6.3).

Landholders Groups

We already heard about the impact of different factors on farmers' income of religious and caste groups, so we went deeper to understand the relationship between land size

TABLE 6.3
Household Income (Rs/Annum) by Social Group (2020–2021)

	Caste									
	SC		ST		OBC		General		Total	
Income Source	Mean	%	Mean	%	Mean	%	Mean	%	Mean	%
Field crops	1,23,714	63	1,52,322	80	2,30,701	77	1,74,124	67	1,91,456	72
MGNREGA	10,406	76	15,487	72	10,762	73	9,279	55	10,749	66
Casual labourer in agriculture	18,304	69	26,273	60	22,283	53	19,211	37	21,310	49
Livestock	27,696	21	22,211	39	25,462	36	27,391	33	25,976	34
Business/self-employment	1,79,423	20	1,48,933	10	2,35,773	23	2,38,442	25	2,28,350	23
Casual labourer in construction	24,038	20	26,976	14	29,353	10	36,396	9	30,525	11
Remittances from migration	28,200	16	22,200	5	27,030	7	28,219	11	27,485	9
Worked as domestic workers	16,000	2	58,000	1	68,800	3	30,077	2	49,094	2
Fruits and vegetables		0		0	2,80,000	1	1,46,667	1	2,13,333	0.2
Sericulture		0		0		0	50,000	0.4	50,000	0.2
Total household income	1,49,085	100	1,78,724	100	2,68,640	100	2,05,149	100	2,22,388	100

Note: % indicates the proportion of households within each category.

TABLE 6.4
Household Income (Rs/Annum) by Social Group (2021–2022)

	Caste									
	SC		ST		OBC		General		Total	
Income Source	Mean	%	Mean	%	Mean	%	Mean	%	Mean	%
MGNREGA	10,509	76	13,564	71	10,410	72	11,271	55	11,069	65
Field crops	1,02,900	41	1,00,961	69	1,16,652	52	1,35,774	49	1,20,926	52
Casual labourer in agriculture	19,027	70	26,616	61	22,990	52	19,396	37	21,804	49
Livestock	27,833	21	22,429	38	25,925	36	27,913	33	26,420	34
Business/self-employment	1,74,704	21	1,49,267	10	2,42,144	23	2,51,231	25	2,36,293	23
Casual labourer in construction/non-agriculture	21,000	20	27,400	14	28,015	10	34,808	9	29,025	11
Remittances from migration (pensions)	28,200	16	22,200	5	27,030	7	28,219	11	27,485	9
Worked as domestic workers	3,000	1	65,000	1	63,867	3	30,077	2	47,806	2
Fruits and vegetables		0		0	3,74,000	1	1,44,000	1	2,71,778	1
Sericulture		0		0		0.4	2,00,000	0.4	2,00,000	0.2
Total household income	1,14,177	100	1,25,436	100	1,56,756	100	1,61,212	100	1,51,431	100

Note: % indicates the proportion of households within each category.

and household income among farmers (Figure 6.4). There was a drastic decrease in the dependency on agriculture and its income in 2021–2022 (Tables 6.5 and 6.6). But there was not much variation in income from livestock and non-agriculture activities

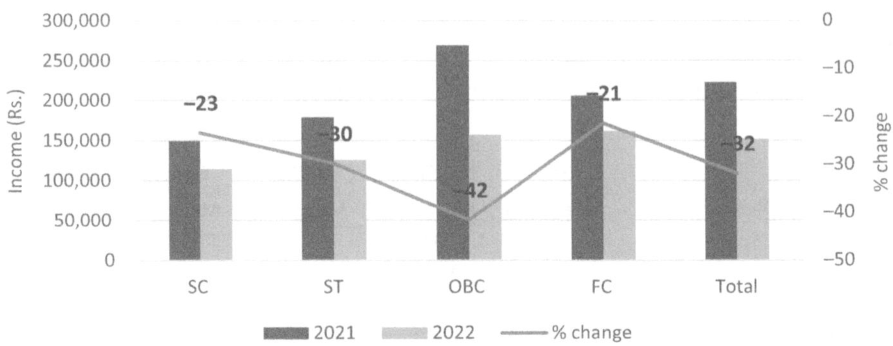

FIGURE 6.3 Income by social group.

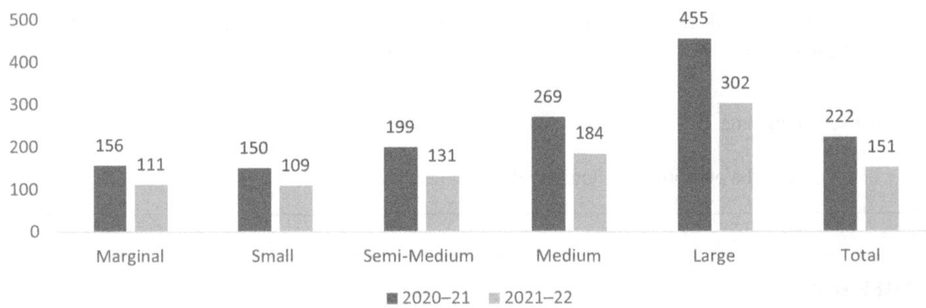

FIGURE 6.4 Household income by land size category (Rs. 1,000).

in all the groups in both the years. We noticed an inverse relationship between farm size and income from casual labour in both agriculture and non-agriculture and farmers who worked as domestic workers. On the contrary, there was a direct relationship between income from crops, livestock, and farm size (Figure 6.5).

Land-Cultivated Category

Households leased in and leased out land to adjust their cultivated area to suit their capability to cultivate. Figure 6.6 depicts the land-cultivated category and income in both normal and abnormal years. Both marginal farmers and large farmers are impacted heavily by the abnormal years, and their incomes were reduced by 36%, while the reduction in income was low among small farmers, which may be due to their multiple sources of income like income from livestock, sheep and goat rearing, dairy and petty business, and causal labour employment.

Poverty Groups

Our findings showed that crop failure had a major impact on the income of both poor and non-poor households, but the non-poor were able to sustain their income through other sources. One surprising finding was that the earnings of poor households from the cultivation of field crops had increased by 44% in 2021–2022 (Tables 6.7 and 6.8).

TABLE 6.5
Household Income (Rs/Annum) by Land-size classification (2020–2021)

	Marginal		Small		Semi-Medium		Medium		Large		Total	
Income Source	**Mean**	**%**	**Mean**	**%**	**Mean**	**%**	**Mean**	**%**	**Mean**	**%**	**Mean**	**%**
Field crops	1,34,241	66	1,11,327	66	1,76,880	72	2,22,197	75	3,96,755	79	1,91,456	72
MGNREGA	10,990	67	10,858	73	11,323	71	10,352	63	6,709	34	10,749	66
Casual labourer in agriculture	22,477	66	22,255	55	21,076	53	20,034	39	19,941	17	21,310	49
Livestock	23,308	25	23,980	31	24,306	33	26,817	40	36,973	37	25,976	34
Business/self-employment	1,41,675	20	2,32,386	17	2,01,500	18	2,59,945	27	2,77,209	43	2,28,350	23
Casual labourer in non-agriculture	34,156	16	32,919	12	23,038	12	35,888	10	26,500	2	30,525	11
Remittances and pensions	28,896	13	28,660	12	27,484	7	26,049	10	24,600	4	27,485	9
Worked as domestic workers	31,400	3	40,500	2	1,02,286	2	35,833	3	12,500	2	49,094	2
Fruits and vegetables	2,80,000	1	70,000	0.2	3,80,000	0.2	70,000	0.2	4,10,000	1	2,13,333	0.2
Sericulture		0		0		0		0	50,000	1	50,000	0.3
Total household income	1,56,219	100	1,49,636	100	1,98,587	100	2,69,288	100	4,54,682	100	2,22,388	100

Note: % indicates the proportion of households within each category.

TABLE 6.6
Household Income (Rs/Annum) by Land size classification (2021–2022)

	Marginal		Small		Semi-Medium		Medium		Large		Total	
Income Source	**Mean**	**%**	**Mean**	**%**	**Mean**	**%**	**Mean**	**%**	**Mean**	**%**	**Mean**	**%**
MGNREGA	10,923	66	10,396	72	13,243	70	9,384	63	7,541	34	11,069	65
Field crops	89,636	48	70,335	46	1,06,016	53	1,38,485	54	2,67,327	58	1,20,926	52
Casual labourer in agriculture	23,135	65	22,823	55	21,616	53	19,879	39	24,118	17	21,804	49
Livestock	24,128	24	23,815	31	24,852	33	27,225	40	37,703	37	26,420	34
Business/ self-employment	1,41,800	20	2,36,364	17	2,14,571	18	2,70,809	27	2,78,256	43	2,36,293	23
Casual labourer in non-agriculture	30,125	16	30,613	12	22,779	11	35,335	10	23,000	2	29,025	11
Remittances and pensions	28,896	13	28,660	12	27,484	7	26,049	10	24,600	4	27,485	9
Worked as domestic workers	31,400	3	40,833	2	90,714	2	38,182	3	12,500	2	47,806	2
Fruits and vegetables	4,50,000	1	55,000	0.2	2,80,000	0	3,34,250	1	1,62,000	2	2,71,778	1
Sericulture		0		0		0		0	2,00,000	1	2,00,000	0.3
Total household income	1,10,821	100	1,09,101	100	1,30,832	100	1,83,518	100	3,01,902	100	1,51,431	100

Note: % indicates the proportion of households within each category.

FIGURE 6.5 Source of income by land size class (% of households).

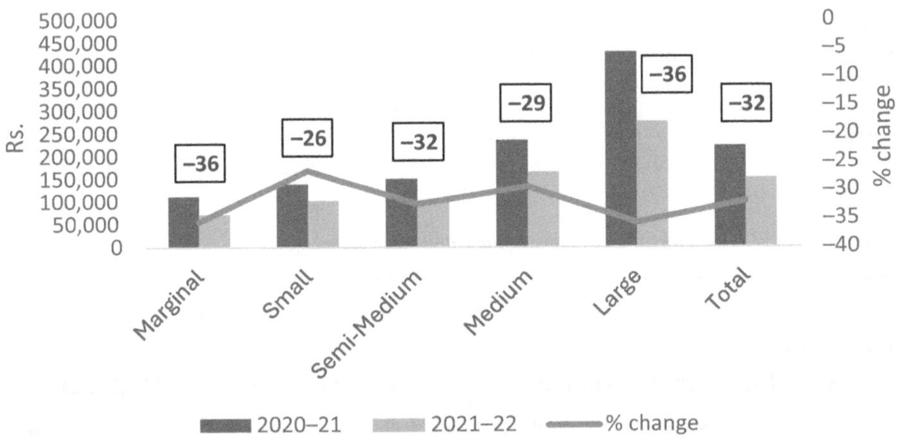

FIGURE 6.6 Household income per annum by land-cultivated class.

Initially, it seemed like a positive change, but, upon further research, it was revealed that the increase was mainly due to higher dependency on non-agricultural sources, which were less impacted by crop loss. For poor households who depended on agriculture, escaping poverty was getting harder each day. This pushes them towards relying more on non-farming sources of income. In 2020–2021, 87% of non-poor households earned income from field crops, but, in the year of the crop failure, this percentage decreased to 57%. As we studied deeper, we found that many of the farmers who were considered poor in the year of the crop failure had been non-poor in the previous year. This highlighted the vulnerability of farmers and their heavy reliance on field crops for income. The crop failure had not only affected their livelihoods but had also pushed them into poverty.

Tenancy Groups

We usually thought that farmers who own their land make more money than the tenants. But our research showed a different story. In both 2020–2021 and 2021–2022, tenant-farmers reported making about the same or even a little more money than

TABLE 6.7
Household Income (Rs/Annum) Sources by Poverty Status in 2020–2021

	2021						2022					
	Non-Poor		Poor		Total		Non-Poor		Poor		Total	
Income Source	**Mean**	**%**	**Mean**	**%**	**Mean**	**%**	**Mean**	**%**	**Mean**	**%**	**Mean**	**%**
MGNREGA	12,284	64	7,859	69	10,749	66	12,496	64	8,404	68	11,069	65
Field crops	2,30,142	87	26,085	41	1,91,456	72	1,47,043	57	46,542	40	1,20,926	52
Casual labourer in agriculture	22,800	49	18,264	48	21,310	49	23,081	48	19,300	50	21,804	49
Livestock	28,204	40	17,875	22	25,976	34	28,624	40	18,275	22	26,420	34
Business/ self-employment	2,42,930	31	30,864	5	2,28,350	23	2,51,408	32	30,864	5	2,36,293	23
Casual labourer in non-agriculture	40,049	10	17,590	14	30,525	11	37,827	9	17,201	14	29,025	11
Remittances pensions	27,685	11	26,640	5	27,485	9	27,685	11	26,640	5	27,485	9
Worked as domestic workers	70,238	2	8,727	2	49,094	2	69,300	2	8,727	2	47,806	2
Fruits and vegetables	2,13,333	1		0	2,13,333	0.3	2,71,778	1		0	2,71,778	1
Sericulture	50,000	0.3		0	50,000	0.3	2,00,000	0.3		0	2,00,000	0.2
Total household income	3,16,019	100	34,333	100	2,22,388	100	2,05,120	100	43,596	100	1,51,431	100

Note: % indicates the proportion of households within each category.

owner-farmers. This was because tenant-farmers were more efficient in using their resources (Ebong et al., 2011). They were careful with their spending and tried to save more money wherever they could because they believed that "Money saved is money earned" (Bansal and Grover, 2019). Instead of buying expensive machinery or hiring labour, they used their own family members to do the work. It not only helped to reduce their cost but also made more profit. Owner-farmers mostly earned from agriculture, public works, livestock, and business/self-employment, whereas the tenant-farmers earned more from casual labour in both agriculture and non-agriculture, remittance, and pension incomes (Table 6.9). The tenant-farmers invested less in purchasing or hiring expensive farm machinery and preferred to provide family labour instead of hiring labour to reduce the cost (Bansal and Grover, 2020). One old farmer told us "*Our daughters can go alone to harvest fodder at any time.*" This showed that it was not just about making money for them, but they loved putting in their effort because farming is their way of life.

However, this was short-lived as the bad year of 2021–2022 hit both groups of farmers hard (Table 6.10). It was a tough time for all households, regardless of whether they owned their land or rented it, and their incomes went down a lot.

TABLE 6.8
Household Income (Rs/Annum) Source by Poverty Status in 2021–2022

	2020–2021						2021–2022					
	Non-Poor		Poor		Total		Non-Poor		Poor		Total	
Income Source	**Mean**	**%**	**Mean**	**%**	**Mean**	**%**	**Mean**	**%**	**Mean**	**%**	**Mean**	**%**
MGNREGA	11,882	64	9,525	68	10,749	66	13,084	64	8,862	66	11,069	65
Field crops	2,17,487	79	1,54,541	63	1,91,456	72	1,56,113	71	26,151	30	1,20,926	52
Casual labourer in agriculture	23,374	50	18,831	47	21,310	49	24,019	50	19,147	47	21,804	49
Livestock	30,186	40	19,035	27	25,976	34	30,655	40	19,323	27	26,420	34
Business/ self-employment	2,52,993	38	27,686	5	2,28,350	23	2,61,902	38	27,029	5	2,36,293	23
Casual labourer in non-agriculture	42,825	10	18,833	12	30,525	11	41,197	10	17,006	12	29,025	11
Remittances and pensions	27,822	12	26,692	6	27,485	9	27,822	12	26,692	6	27,485	9
Worked as domestic workers	79,471	2	14,667	2	49,094	2	78,875	2	14,667	2	47,806	2
Fruits and vegetables	2,38,000	1	90,000	0.2	2,13,333	0.2	2,71,778	1		0	2,71,778	1
Sericulture	50,000	0.2		0	50,000	0.2	2,00,000	0.2		0	2,00,000	0.2
Total household income	3,09,363	100	1,23,643	100	2,22,388	100	2,55,457	100	33,326	100	1,51,431	100

Note: % indicates the proportion of households within each category.

TABLE 6.9
Household Income (Rs/Annum) by Land Tenancy Status (2020–2021)

	Owner-cultivator		Tenant-cultivator		Total	
Source of Income	**Mean**	**%**	**Mean**	**%**	**Mean**	**%**
Field crops	1,90,422	72	2,13,802	63	1,91,456	72
MGNREGA	10,822	66	9,205	58	10,749	66
Casual labourer in agriculture	21,117	48	23,957	65	21,310	49
Livestock	25,906	34	27,778	25	25,976	34
Business/self-employment	2,28,987	23	2,14,429	19	2,28,350	23
Casual labourer in non-agriculture	29,660	11	39,429	19	30,525	11
Remittances from migration (pensions)	27,426	9	28,200	14	27,485	9
Worked as domestic workers	52,828	2	13,000	4	49,094	2
Fruits and vegetables	2,00,000	0.3	2,80,000	1	2,13,333	0.3
Sericulture	50,000	0.3		0	50,000	0.2
Total household income	2,22,554	100	2,19,288	100	2,22,388	100

Note: % indicates the proportion of households within each category.

TABLE 6.10
Household Income (Rs/Annum) by Land Tenancy Status (2021–2022)

Income Source	Owner-cultivator		Tenant-cultivator		Total	
	Mean	%	Mean	%	Mean	%
MGNREGA	11,143	66	9,473	57	11,069	65
Field crops	1,19,319	52	1,58,687	42	1,20,926	52
Casual labourer in agriculture	21,538	48	25,543	64	21,804	49
Livestock	26,332	34	28,824	24	26,420	34
Business/self-employment	2,37,225	23	2,15,857	19	2,36,293	23
Casual labourer in construction/non-agriculture	28,391	11	35,500	19	29,025	11
Remittances from migration (pensions)	27,426	9	28,200	14	27,485	9
Worked as domestic workers	51,536	2	13,000	4	47,806	2
Fruits and vegetables	2,49,500	1	4,50,000	1	2,71,778	1
Sericulture	2,00,000	0.4		0	2,00,000	0.2
Total household income	1,51,282	100	1,54,222	100	1,51,431	100

Note: % indicates the proportion of households within each category.

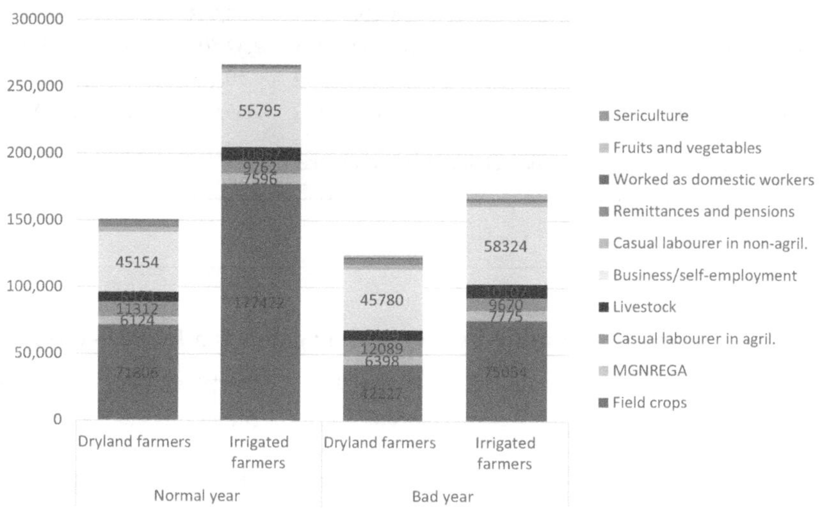

FIGURE 6.7 Household incomes by dryland and irrigated farmers (Rs.).

Irrigation Groups

Dryland farmers could not grow diversified crops due to water scarcity, which meant they were not getting as much yield from their crops. Because of this, the prices of crops have gone up, and, as a result, they were losing money (Samuel et al., 2022). In contrast, irrigated farmers enjoyed significantly higher household incomes due to their ability to cultivate a wider range of crops. This was a clear indication that irrigation not only increased agricultural incomes but also opened opportunities for earning from other sources (Figure 6.7). The income from public works, casual labour, self-employment in non-agriculture, and livestock was also higher for irrigated farmers, in both normal and bad years (Tables 6.11 and 6.12). On the contrary, dryland farmers realize their dependency on non-farm incomes and depend much on family members employed elsewhere like dryland farmers (Oni et al., 2011).

TABLE 6.11
Household Income (Rs/Annum) by Dry and Irrigated Land (2020–2021)

	Dry Land		Irrigated Land		Total	
Source of Income	**Mean**	**%**	**Mean**	**%**	**Mean**	**%**
Field crops	1,17,714	61	2,27,464	78	1,91,456	72
MGNREGA	8,625	71	12,252	62	10,749	66
Casual labourer in agriculture	20,200	56	22,186	44	21,310	49
Livestock	24,904	28	26,465	38	25,976	34
Business/self-employment	2,05,246	22	2,42,586	23	2,28,350	23
Casual labourer in construction/non-agriculture	25,177	15	35,873	9	30,525	11
Remittances from migration (pensions)	28,435	15	26,042	6	27,485	9
Worked as domestic workers	43,059	3	55,933	2	49,094	2
Fruits and vegetables	1,75,000	0.3	2,32,500	0.3	2,13,333	0.2
Sericulture		0	50,000	0.3	50,000	0.2
Total household income	1,52,175	100	2,65,991	100	2,22,388	100

Note: % indicates the proportion of households within each category.

TABLE 6.12
Household Income (Rs/Annum) by Dry and Irrigated Land (2021–2022)

	Dry Land		Irrigated Land		Total	
Income Source	**Mean**	**%**	**Mean**	**%**	**Mean**	**%**
MGNREGA	9,011	71	12,540	62	11,069	65
Field crops	98,203	43	1,31,673	57	1,20,926	52
Casual labourer in agriculture	21,587	56	21,977	44	21,804	49
Livestock	26,033	28	26,597	38	26,420	34
Business/self-employment	2,08,090	22	2,53,583	23	2,36,293	23
Casual labourer in non-agriculture	23,638	15	34,481	9	29,025	11
Remittances from migration (pensions)	28,435	15	26,042	6	27,486	9
Worked as domestic workers	43,353	3	53,214	2	47,807	2
Fruits and vegetables	1,44,000	1	3,74,000	1	2,71,778	1
Sericulture		0	2,00,000	0.3	2,00,000	0.2
Total household income	1,24,960	100	1,67,869	100	1,51,431	100

Note: % indicates the proportion of households within each category.

We could realize the multiplier effect of irrigation on the rural economy. It was not just limited to the agricultural sector, but it also had a ripple effect on all other sectors. The increase in irrigation facilities created a domino effect, leading to a rise in income and employment opportunities for the farmers and the community.

COMPARISONS AMONG STATES

As our team gathered together and sat in a place, our minds wandered back to the data we had compiled. Telangana had the highest average income of over Rs. 2.7 lakh, but, the following year, things took a turn for the worse (Figure 6.8). Due to unforeseen circumstances, the incomes dropped drastically in all three states. It was a significant blow to the households, and our team was determined to find out the reasons behind this drastic decrease.

When we dove into the data again, we noticed that, in a normal year, Telangana households earned more income from all sources except for livestock, while Karnataka fared better (Table 6.13). Andhra Pradesh, on the contrary, performed better in terms of remittances and pensions.

But, in the bad year, the situation was quite different. In the face of adversity, Andhra Pradesh managed to take care of its citizens' well-being. Even though Telangana households still earned more income from sources like public works, casual labour in agriculture and non-agriculture, and self-employment in non-agriculture, it was Andhra Pradesh that stood out by its better and universal reach of pensions (Table 6.14).

Our research opened our eyes to the diverse income sources of farmers and the challenges they face in sustaining their livelihoods. We saw the highs and lows of economic prosperity, from good harvests to crop failures, and how these sources of income provided a safety net for the farming community. We realized that relying solely on agriculture could be risky, especially with the changing weather.

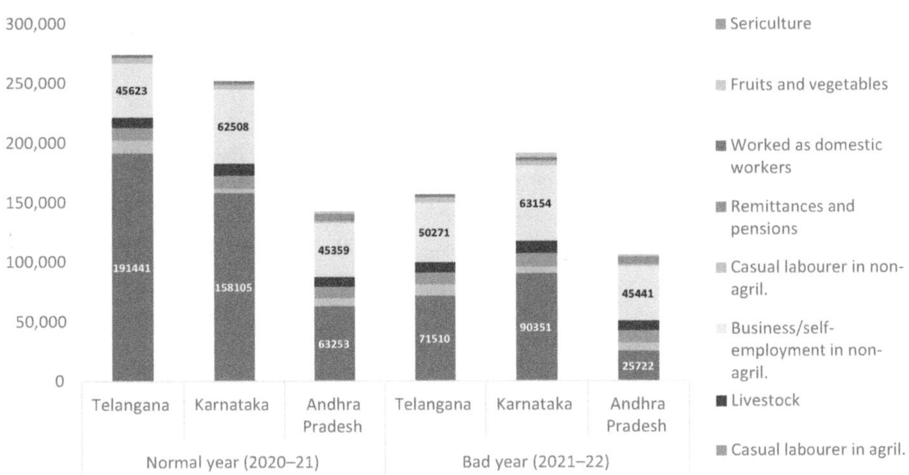

FIGURE 6.8 Household income by state (in Rs.)

TABLE 6.13
Household Income by State (2020–2021)

	State Name							
	Telangana		Karnataka		Andhra Pradesh		Total	
Income Source	**Mean**	%	**Mean**	%	**Mean**	%	**Mean**	%
Field crops	2,20,047	87	2,10,806	75	1,19,346	53	1,91,456	72
MGNREGA	15,459	72	7,246	53	8,803	73	10,749	66
Casual labourer in agriculture	23,801	44	21,352	51	19,223	51	21,310	49
Livestock	21,611	40	29,797	33	27,399	29	25,976	34
Business/self-employment	2,68,372	17	2,50,033	25	1,81,434	25	2,28,350	23
Casual labourer in non-agriculture	58,263	8	25,715	15	15,781	10	30,525	11
Remittances and pensions	36,000	1	19,291	7	30,000	20	27,485	9
Worked as domestic workers	95,750	2	51,357	3	8,600	2	49,094	2
Fruits and vegetables		0	2,35,000	0.3	2,02,500	1	2,13,333	0.2
Sericulture		0	50,000	0.3		0	50,000	0.2
Total household income	2,74,129	100	2,52,156	100	1,43,035	100	2,22,388	100

Note: % indicates the proportion of households within each category.

TABLE 6.14
Household Income by State (2021–2022)

	State Name							
	Telangana		Karnataka		Andhra Pradesh		Total	
Income Source	**Mean**	%	**Mean**	%	**Mean**	%	**Mean**	%
MGNREGA	13,796	71	10,491	53	8,947	73	11,069	65
Field crops	1,19,183	60	1,39,002	65	85,740	30	1,20,926	52
Casual labourer in agriculture	23,233	43	22,756	51	19,724	52	21,804	49
Livestock	21,062	40	30,887	33	28,465	29	26,420	34
Business/self-employment	2,95,709	17	2,52,617	25	1,81,762	25	2,36,293	23
Casual labourer in non-agriculture	54,622	8	24,506	15	16,073	10	29,025	11
Remittances and pensions	36,000	1	19,291	7	30,000	20	27,485	9
Worked as domestic workers	94,000	2	52,714	3	8,600	2	47,806	2
Fruits and vegetables		0	3,95,000	1	1,73,200	1	2,71,778	1
Sericulture		0	2,00,000	0.2		0	2,00,000	0.1
Total household income	1,57,223	100	1,90,706	100	1,06,604	100	1,51,431	100

Note: % indicates the proportion of households within each category.

REFERENCES

Bansal S, Grover DK. (2019). Benefit Cost analysis of Tenant Farmers. A Case Study of Sugarcane Cultivation. *Journal of Agricultural Development and Policy*, 29(2), pp. 155–161.

Bansal S, Grover DK. (2020). Cost and Return Structure of Wheat Crop under Tenant Farmers in Punjab: A Field Survey. *Indian Journal of Economics and Development*, 16(2S), pp. 500–503.

Barrett CB, Reardon T, Webb P. (2001). Nonfarm Income Diversification and Household Livelihood Strategies in Rural Africa: Concepts, Dynamics, and Policy Implications. *Food Policy*, 26(4), pp. 315–331.

Ebong VO, Effiong EO, Eshiet AJ, Nuka H. (2011). Resource Use Efficiency of Land Owner and Tenants in Cassava Based Farms in Akwa Ibom State, Nigeria: A Comparative Analysis. *Agriculture and Biology Journal of North America*, 2, pp. 1048–1052.

Ellis F. (2000). *Rural Livelihoods and Diversity in Developing Countries*. Oxford University Press.

Hazell P, Rahman A. (2014). *New Directions for Smallholder Agriculture*. Oxford University Press.

Jayne TS, Mather D, Mghenyi E. (2010). Principal Challenges Confronting Smallholder Agriculture in Sub-Saharan Africa. *World Development*, 38(10), pp. 1384–1398.

Narayanamoorthy A. (2017). Farm Income in India: Myths and Realities. *Indian Journal of Agricultural Economics*, 72(1), pp. 49–75.

Nwibo SU, Nwakpu C, Eze AV. (2017). Income and Investment Portfolio of Smallholder Farmers in Ezza South Local Government Area of Ebonyi State, Nigeria. *European Journal of Scientific Research*, 145(3), pp. 328–335.

Odoh NE, Nwibo SU, Eze A, Igwe EO. (2020). Farmers Income and Savings Pattern in Benue State, Nigeria. *Journal of Agricultural Extension*, 24(1), pp. 128–137.

Olawepo RA. (2010). Determining Rural Farmers' Income: A Rural Nigeria Experience. *Journal of African Studies and Development*, 2(4), pp. 99–108.

Oni SA, Maliwichi LL, Obadire OS. (2011). Assessing the Contribution of Smallholder Irrigation to Household Food Security, in Comparison to Dryland Farming in Vhembe District of Limpopo Province, South Africa. *African Journal of Agricultural Research*, 6(10), pp. 2188–2197.

Reardon T, Berdegué J, Escobar G. (2001). Rural Nonfarm Employment and Incomes in Latin America: Overview and Policy Implications. *World Development*, 29(3), pp. 395–409.

Samuel J, Rao CAR, Raju BMK, Reddy AA, Pushpanjali, Reddy AGK, Kumar RN, Osman M, Singh VK, Prasad JVNS. (2022). Assessing the Impact of Climate Resilient Technologies in Minimizing Drought Impacts on Farm Incomes in Drylands. *Sustainability*, 14(1), p. 382.

Sen A, Bhatia MS. (2004). *State of the Indian Farmer: A Millenium Study 14. Cost of Cultivation and Farm Income in India*. Academic Foundation, New Delhi.

7 Patterns in Household Savings

In the previous chapter, we explored how economic fluctuations, coupled with climatic risks, have led to significant income variability for farming households (Dercon, 2005; Carter and Barrett, 2006; Daniel, 2019). In such circumstances, household savings become essential to protect against these shocks and ensure financial resilience (Deaton, 1992; Abdelkhalek et al., 2010). For farmers seeking to expand their operations and increase their earnings, effective management of savings is crucial (Zeller and Sharma, 1998; Arifin, 2018). Unfortunately, the farming sector offers limited opportunities for saving, a challenge that is exacerbated by increasingly unpredictable weather patterns (Rosenzweig & Binswanger, 1993). As a result, many farmers find it difficult to escape the cycle of poverty (Lipton, 1982; Ellis, 1998; Albu and Scott, 2001; Anderson et al., 2017).

Still, we heard many fascinating stories about how farmers in developing countries like India were able to build their savings with modest incomes, so we decided to unveil the secrets behind their success. The farmers appreciated our incredible curiosity, and they generously started sharing their stories. They explained that, to increase their savings, they grew multiple crops instead of just one. This helped them diversify their income sources and reduce the risk of crop failure. We already heard from them that many farmers engaged in non-farm activities, which helped to supplement their income. We were truly impressed by their determination and resilience in finding ways to increase their savings.

After questioning for a long time, one of the key factors we discovered was their participation in various government schemes. These schemes provided them with financial support and resources to improve their farming methods and increase their productivity. It was heartening to see how these schemes were making a positive impact on their lives. But the greatest secret we uncovered was the magic of investing in assets like land, livestock, and equipment. These assets provided a steady source of income and helped to increase their net worth over time. We were impressed by their long-term thinking and planning. Through our research, we also learned that farmers allocate a portion of their yearly income to old age and invest in long-term schemes. Also, some of them try to save as a matter of habit, and saving is usually for precautionary reasons (Isiorhovoja et al., 2020). This showed us that savings were not just a chore for them but a journey of hope and determination for a better future. They were not only saving for themselves but for their future generations as well.

DIVERSE PATH TO SAVINGS

Our research on farmers in Telangana, Karnataka, and Andhra Pradesh has taught us a lot. As our team inquired more into the financial habits of the farmers, they said:

DOI: 10.1201/9781003582878-7

> We work hard, but it is tough to set aside anything for savings. Our income is so unpredictable that we are always struggling to get it. Every rupee counts for us.

Many of the villages did not have banks or financial organizations nearby to keep whatever they managed to save. It made their situation more challenging.

But, amidst this struggle, the farmers found their paths whether it was through self-help groups (SHGs), co-operatives, farmer producer organizations (FPOs), or some other means. We never imagined that their savings would be so diverse and reflective of their unique circumstances. Among these, SHG was a very reliable path for the hardy farmers. "*Our lives have somehow improved since we started the SHGs. It's not just about saving money, it's about investing for a better future,*" said Bhubnesh Kumar, a small farmer and a member of an SHG in Telangana. In the villages in Telangana, the majority of the farmers depend on rainfed agriculture as the area is prone to more frequent and longer dry spells, drought, and erratic rainfall. "*To address the challenges, we formed these groups to save money collectively. These enable us to buy essential materials like hybrid seeds and fertilizer, and sometimes meet emergencies,*" added Kumar.

These SHGs offered farmers the opportunity to save a portion of their earnings to secure their future. But the savings pattern was also different across the regions and groups. We saw that approximately half of the farmers in Telangana used SHGs to save money. However, the situation was quite different in Karnataka. Farmers here rarely used SHGs to save money, and even saving was not a common practice in this state. But, in Andhra Pradesh, the saving habits of farmers were somewhere between Telangana and Karnataka where some farmers preferred only SHGs to save their money. This made us wonder – why the difference? Upon inspection we found out that, in Telangana and Andhra Pradesh, the governments encouraged farmers to save with SHGs, while, in Karnataka, they did not focus on it as much.

PROBING FURTHER INTO THE GROUPS

Religious Groups

Since religious beliefs strongly influence how people save and invest their money (León, 2013), we did more research to understand how religion affects farmers' saving habits. It would help us to see why different cultures have different economic outcomes (Inglehart and Welzel, 2010). After talking extensively with the farmers, we found that Hindus and Muslims tended to have similar savings patterns, choosing different methods for saving their money. However, Christians had a slightly different approach; they mostly relied on SHGs (Table 7.1). Surprisingly, the average savings were quite similar among Hindu and Christian farmers about Rs 9,500, but it was notably less among Muslim farmers. But these savings were still not enough if we think about the farming costs.

Social Groups

We found some pieces of evidence that show how a specific caste can directly impact the savings behaviour of a household (Bonte and Filipiak, 2011). This social stratum was very common in villages and can make it hard for people to access different resources (Shah et al., 2006). Despite facing numerous obstacles such as low income,

TABLE 7.1
Savings (in Rs.) by Religion Category

	Religion							
	Hindu		Muslim		Christian		Total	
Type of Institution	**Mean**	**%**	**Mean**	**%**	**Mean**	**%**	**Mean**	**%**
Self-help group amount saved (cumulative)	29,857	23	32,550	15	22,236	43	29,690	23
Co-operatives amount saved	87,313	2	68,500	4		0	86,206	2
Farmer producer organization	24,733	1		0		0	24,733	1
Others	2,500	0.3	0	2		0	1,250	0.2
Total savings	9,323	100	7,642	100	9,668	100	9,267	100

Note: % indicates the proportion of households within each category.

limited resources, social discrimination, high levels of debt, limited access to formal banking systems, agricultural risks, and lack of financial literacy, these farmers showed they could change. They were determined to save money for a better future. We were amazed to see that 23% of farmers from marginalized communities were wise savers, especially those from scheduled caste (SC) households (Table 7.2). It was clear that these farmers understood the importance of saving and were actively participating in SHGs to achieve their financial goals. The government initiatives also offered a ray of hope for these farmers. Government schemes like Pradhan Mantri Fasal Bima Yojana and Pradhan Mantri Kisan Samman Nidhi scheme offered financial support for SC, scheduled tribe, and other backward class farmers. Also, organizations like the National Scheduled Castes Finance and Development Corporation, National Backward Classes Finance and Development Corporation, and National Scheduled Tribes Finance and Development Corporation supported them with income-generating activities, including agriculture.

TABLE 7.2
Patterns of Savings (in Rs.) by Households by Social Group

	Caste									
	Scheduled Castes		Scheduled Tribes		Other Backward Classes		Forward Caste		Total	
Savings Channels	**Mean**	**%**	**Mean**	**%**	**Mean**	**%**	**Mean**	**%**	**Mean**	**%**
Self-help group	34,098	25	29,490	32	38,618	22	20,186	22	29,690	23
Co-operatives		0		0	1,00,143	4	63,692	2	86,206	2
Farmer producer organizations	60,000	1		0	23,846	2	1,000	0.3	24,733	1
Others		0		0	0	0	2,500	0.3	1,250	0.1
Total savings	9,047	100	9,291	100	12,608	100	6,077	100	9,267	100

Note: % indicates the proportion of households within each category.

Landholding Groups (Own Land)

When we focused on how the farmers' land ownership category affected their ability to save money, we noticed that a higher number of small and medium farmers had savings with SHGs compared to both marginal and large farmers (Table 7.3). In fact, the share of large farmers who saved with SHGs was drastically lower compared to other land-size groups. However, when it came to the amount of savings, large farmers saved significantly more than others in SHGs. But that was not all; the households that had savings with co-operatives were higher among marginal and large farmers, which was the exact opposite pattern of savings with SHGs. In general, savings amounts in co-operatives were much higher compared to SHGs. Many medium and large farmers were also saving with commercial banks and in larger quantities.

Landholding Groups (Total Cultivated Land)

Another aspect that caught our attention was the difference in savings among farmers based on the size of their cultivated land, where the savings with SHGs were significantly higher among marginal farmers compared to large farmers (Table 7.4). So, we were determined to find out why. When we further interviewed the farmers they told us:

TABLE 7.3
Household Savings (in Rs.) by Own Land Size Category

Savings Channels	Marginal		Small		Semi-Medium		Medium		Large		Total	
	Mean	%	Mean	%	Mean	%	Mean	%	Mean	%	Mean	%
Self-help group	29,143	21	29,561	30	28,799	26	29,274	21	58,800	6	29,143	21
Co-operatives	62,000	3	63,667	2	1,22,727	2	71,556	2	81,667	3	62,000	3
Farmer producer organizations		0	60,000	0.3	30,100	2	2,500	1		0		0
Others		0		0	0	0	2,500	0		0		0
Total savings	7,709	100	10,544	100	11,175	100	7,748	100	6,612	100	7,709	100

Note: % indicates the proportion of households within each category.

TABLE 7.4
Household Savings (in Rs.) by Land Cultivated Size Category

	Marginal		Small		Semi-Medium		Medium		Large		Total	
Savings Channels	Mean	%	Mean	%	Mean	%	Mean	%	Mean	%	Mean	%
Self-help group	41,656	25	30,880	25	26,158	27	27,888	23	38,834	14	29,690	23
Co-operatives	76,667	8	43,500	1	1,32,000	2	71,400	2	68,857	3	86,206	2
Farmer producer organizations		0	60,000	1	30,100	2	2,500	1		0	24,733	1
Other		0		0	0	0		0	2,500	0.3	1,250	0.2
Total savings	16,164	100	8,514	100	10,793	100	7,769	100	8,110	100	9,267	100

Note: % indicates the proportion of households within each category.

> We collect our savings and take small loans from the group to invest in our farms. This not only boosts our yields but also makes us feel more secure.

Poverty Groups

Our team spent countless hours interviewing farmers in different villages. What stood out was that non-poor farmers had a more diverse income source, had a better access to financial services, and could save and invest in their farms. Surprisingly, many from both groups saved with SHGs and FPOs, even during distressing years (Table 7.5). It was nice to see that despite their struggles, more poor farmers prioritized saving, seeing it as a social obligation. Not only this, there was a growing trend of savings with FPOs, indicating their importance as a saving institution for farmers, including the poor.

Tenancy Groups

One trend that we noticed every time in our research is that tenant farmers face numerous obstacles in every aspect even in saving money. The biggest hurdle was not owning land, which created difficulties for the farmers to access financial services such as co-operatives and SHGs (Table 7.6). To understand this better, we requested a farmer to explain in detail. He said:

> We are aware of the benefits of saving with cooperatives, but we simply do not have the means to do so. Without land ownership, we cannot access loans.

Irrigation Groups

Irrigated farmers admitted that they saved terribly more compared to their dryland counterparts. This led us to investigate further, and what we discovered was concerning. Only 21% of dryland farmers saved with SHGs, and merely 1% saved with co-operatives (Table 7.7). In contrast, 25% of irrigated farmers saved with SHGs, and their savings amounted to three times those of dryland farmers. Unlike irrigated

TABLE 7.5
Household Savings (in Rs.) by Poverty Status

Savings Channels	2020–2021						2021–2022					
	Non-Poor		Poor		Total		Non-Poor		Poor		Total	
	Mean	%	Mean	%	Mean	%	Mean	%	Mean	%	Mean	%
Self-help group	37,179	22	17,369	26	29,690	23	39,203	20	21,474	26	29,690	23
Co-operatives	94,188	2	79,111	4	86,206	2	61,583	2	99,636	3	86,206	2
Farmer producer organizations	29,900	1	14,400	1	24,733	1	36,625	1	11,143	1	24,733	1
Others	2,500	0.2	0	0	1,250	0.2	1,250	0.2		0	1,250	0.2
Total savings	10,030	100	7,733	100	9,267	100	9,270	100	9,262	100	9,267	100

Note: % indicates the proportion of households within each category.

TABLE 7.6
Households Savings (in Rs.) by Land Tenancy Status, 2022

	Owner-cultivator		Tenant-cultivator		Total	
Savings Channels	**Mean**	**%**	**Mean**	**%**	**Mean**	**%**
Self-help group	29,715	23	29,042	17	29,690	23
Co-operatives	86,206	3		0	86,206	2
Farmer producer organizations	24,733	1		0	24,733	1
Others	1,250	0.2		0	1,250	0.2
Total savings	9,503	100	4,840	100	9,267	100

Note: % indicates the proportion of households within each category.

TABLE 7.7
Household Savings (in Rs.) by Irrigated and Dryland Farmers

	Dry Land		Irrigated Land		Total	
Savings Channels	**Mean**	**%**	**Mean**	**%**	**Mean**	**%**
Self-help group	13,626	21	38,019	25	29,690	23
Co-operatives	47,200	1	92,931	3	86,206	2
Farmer producer organizations		0	24,733	2	24,733	1
Others		0	1,250	0.2	1,250	0.2
Total savings	3,239	100	13,010	100	9,267	100

Note: % indicates the proportion of households within each category.

farmers, dryland farmers depend solely on unpredictable rainfall. The higher input costs for water and other resources resulted in lower profits and added financial burden for dryland farmers, leaving them with little to no savings.

COMPARISONS AMONG STATES

We soon discovered that the savings behaviour of farmers was influenced by factors like the local economy, credit availability, risks, and cultural norms. In areas with high living costs or limited credit, farmers tend to save more for emergencies or farm investments. Social norms, like saving for future generations, also played a role. We found big differences in savings rates among states. Telangana and Andhra Pradesh have more savings through SHGs because their governments actively support these groups, unlike in Karnataka (Table 7.8). Moreover, farmers in Telangana also save through co-operatives and FPOs, which was not the case in other states. This shows that different regions have different ways of saving money.

After listening intently as the farmers shared their stories, we felt a sense of frustration at the unfairness of the situation. These hard-working farmers were being denied the opportunity to secure their future. But we also discovered that participating in

TABLE 7.8
Household Savings (in Rs.) by States

	State Name							
	Telangana		Karnataka		Andhra Pradesh		Total	
Savings Channels	**Mean**	**%**	**Mean**	**%**	**Mean**	**%**	**Mean**	**%**
Self-help group	45,320	46	17,833	1	283	23	29,690	23
Co-operatives	86,206	7		0		0	86,206	2
Farmer producer organizations	26,429	3	1,000	0.3		0	24,733	1
Others		0	0	0	2,500	0.3	1,250	0.2
Total savings	28,162	100	343	100	83	100	9,267	100

Note: % indicates the proportion of households within each category.

financial literacy programmes and developing a household budget can be effective ways for small farmers to increase their savings. These programmes helped farmers understand financial management and identify areas where they could reduce costs and increase savings.

REFERENCES

Abdelkhalek T, Arestoff F, de Freitas NM, Mage S. (2010). A Micro Econometric Analysis of Households Saving Determinants in Morocco. *African Review of Money Finance and Banking* (Supplementary Issues of S & D), pp. 7–27.

Albu M, Scott A. (2001). *Understanding Livelihoods That Involve Micro-Enterprise: Markets, and Technological Capacities in the SL Framework, Intermediate Technology Development Ground War Wickshire*. Intermediate Technology Development Group Practical Answers to Poverty.

Anderson J Ma rita C, Musiime D, Thiam M. (2017). National Survey and Segmentation of Smallholder Households in Nigeria: Understanding Their Demand for Financial, Agricultural, and Digital Solutions. CGAP Working Paper. p. 89.

Arifin AZ. (2018). Influence Factors toward Financial Satisfaction with Financial Behavior as Intervening Variable on Jakarta Area Workforce. *European Research Journal*, 21(1), pp. 90–103.

Bonte W, Filipiak W. (2011). Financial Investments, Information Flows, and Caste Affiliation: Empirical Evidence from India. *Schumpeter Discussion Papers No. 2011–2014.*

Carter MR, Barrett CB. (2006). The Economics of Poverty Traps and Persistent Poverty: An Asset-Based Approach. *The Journal of Development Studies*, 42(2), pp. 178–199.

Daniel A-M. (2019). Determinants of Rural Household Savings Behaviour: The Case of Tomato Farmer in Ghana. *Review of Agricultural and Applied Economics*, 22(2), pp. 55–70.

Deaton A. (1992). *Understanding Consumption*. Oxford University Press.

Dercon S. (2005). Risk, Insurance, and Poverty: A Review. In S. Dercon (Ed.), *Insurance against Poverty* (pp. 9–37). Oxford University Press.

Ellis F. (1998). Household Strategies and Rural Livelihood Diversification. *Journal of Development Studies*, 35(1), pp. 1–38.

Inglehart R, Welzel C. (2010). Changing Mass Priorities: The Link between Modernization and Democracy. *Perspectives on Politics*, 8(2), pp. 551–567.

Isiorhovoja RA, Ebewore SO, Nwachi S. (2020). Farmers' Attitude and Behavior toward Savings in Ika South Local Government Area of Delta State, Nigeria. *Asian Journal of Agriculture and Rural Development*, 10(1), pp. 406–419.

León AK. (2013). Religion and Economic Outcomes – Household Savings Behavior in the USA. *Working Paper Series in Economics*. University of Lüneburg, No. 268.

Lipton M. (1982). *Why Poor People Stay Poor*. In Harriss J. (ed.), *Rural Development: Theories of Peasant Economy and Agrarian Change* (1st ed.). Routledge. https://doi.org/10.4324/9781003431763

Rosenzweig MR, Binswanger HP. (1993). Wealth, Weather Risk and the Composition and Profitability of Agricultural Investments. *The Economic Journal*, 103(416), pp. 56–78, https://doi.org/10.2307/2234337

Shah G, Mander H, Thora S, Deshpande S, Baviskar A. (2006). *Untouchability in Rural India*. SAGE.

Zeller M, Sharma M. (1998). *Rural Finance and Poverty Alleviation*. International Food Policy Research Institute (IFPRI).

8 Debt Burden of the Households

The Indian peasant is born in debt, lives in debt and dies in debt.

Darling (1925)

After spending hours researching and talking to farmers, we felt like we understood their lives pretty well. But then we realized that a major issue behind this that was causing them a lot of stress and difficulty was "debt," and also we came to understand the truth behind the above statement by M. L. Darling. All our team members had heard about this problem earlier, but we never realized how serious it was until we saw it ourselves. Changes in rainfall, frequent drought, floods, and high temperatures reduced the productivity of major crops like wheat, rice, and maize. It affected the farmer's lives by reducing their income and leading to more debt of the farmers. Further, because of the high cultivation cost, farmers were not able to make as much money in return (Sajjad and Chauhan, 2012). No matter how much they earned, it was often difficult for them to save even a little. So, many farmers depend on loans from formal and informal sources as loans are very important to meet their farming expenses (Pandey, 2016). However, when they faced any unforeseen challenges like bad weather or low crop harvests, they found it difficult to repay their loans. So, they would take out new loans to pay off the previous ones and trap themselves in a never-ending cycle of debt. Mainly crop failure, inadequate irrigation facilities, and the lack of marketing facilities were the reasons behind this situation (Lavanya, 2016).

We can remember one farmer who shared his story with us. He borrowed some money from a formal organization to buy some farming inputs and accessories. But a sudden calamity destroyed his entire crop and made him destitute. With no way to repay the loan, he had to borrow another loan to settle the first one. This cycle repeated for years, leaving him heavily in debt and financial struggles. Unfortunately, it was a common story for almost every household. A huge number of farming households owed an average of about Rs. 74,000. More than half of the households were stuck in a debt cycle, with the hardest hit areas being Andhra Pradesh, Telangana, Kerala, and Karnataka (Reddy et al., 2020), and the number of indebted farmers had risen from 25% of the total rural households in India in 1992 to about 46% in 2013 (NSSO, 2014).

This story also reminded us of another incident that ran in the news a few years back. On a humid day in September 2018, in Karnataka's Mandya district, 37-year-old farmer Nandish and his family were supposed to have dinner together. He was burdened by Rs. 20 lakh loans from various sources, saw his crops fail, and couldn't repay the debts. That night, he poisoned the food they were to eat. In a note, he wrote, "*I am not dying as a coward, but I could not clear my liabilities despite my*

DOI: 10.1201/9781003582878-8

best efforts." Nandish was one of 2,000+ farmers who died by suicide in Karnataka in 2018 (Kumar, 2018).

Although there was not a single cause for these suicides, the growing personal debt poses a significant threat to many farmers. As we spent time with these farmers, heard their stories, and recalled several tragic news in recent years, we found that, even though commercial banks were the primary source of credit for farmers in all three states in 2020, informal sources like money lenders and friends or relatives also played a significant role. Even the informal sources were preferable to the farmers (Narayanan, 2016). But, as the year 2021 began, the crop disaster, as well as the pandemic, affected the farming sector, and there was a change in the indebtedness patterns, with an increase in the share of informal lending sources, showing that farmers were forced to take loans at higher interest rates as they are losing income sources and trapped in financial difficulties.

This chapter focuses on the debt burden of the farmers. The debt burden on farmers in India and other developing countries has been a significant concern, exacerbated by factors like declining agricultural profitability, crop failures, and limited access to formal credit. In India, small and marginal farmers often resort to informal sources of credit, such as money lenders, due to the inaccessibility of formal banking systems. This reliance on high-interest loans traps farmers in a cycle of debt, contributing to distress and, in extreme cases, suicides (Isakson, 2016; Reddy and Mishra, 2012). Similar patterns are observed in other developing countries, where agricultural finance is often inadequate, leading to widespread indebtedness. In sub-Saharan Africa, the absence of robust financial institutions forces farmers to rely on informal credit, often at predatory rates (Mellor, 2017; Chavas et al., 2019). In Southeast Asia, particularly in Vietnam and the Philippines, informal lending networks continue to dominate rural economies, further deepening the debt crisis among farmers (Hoang et al., 2023; ADB, 2020). The COVID-19 pandemic has intensified these challenges, as disruptions in agricultural markets and supply chains have compounded financial difficulties for farmers across the developing world (Dev and Sengupta, 2020; FAO, 2021). Efforts to alleviate this burden require comprehensive reforms in agricultural finance, enhanced access to formal credit, and supportive policies to ensure the financial sustainability of smallholder farmers (Binswanger-Mkhize, 2012; Torhonen et al., 2019).

Sajjad and Chauhan (2012) observed that the rising cost of cultivation and diminishing net returns are most significant for the incidences of indebtedness. The farmers are not getting the appropriate price for their crops due to the present agricultural credit system.

DEBT DYNAMICS

Religious Groups

The debt situation affected all groups, regardless of their religious or other identities. The need for money often exceeds the lender's preference, forcing farmers to make difficult decisions. In India, people of all faiths, Hindus, Muslims, Sikhs, Christians, and others, heavily depend on farming for their income. Despite this, when we looked at an earlier survey, we found that Sikh households (near about 70%) had the highest percentage of debt (NSS, 2013). We became curious, why was there a difference? This was not just about religion. There had to be something more. Factors like land ownership, access to credit, and other socio-economic factors also played a role in farmers' borrowing pattern. But why were Sikhs more indebted than others? When we delved into our research, conducting interviews with farmers from different religious groups, we easily found the answer. It was not only that Sikhs had easier access to credit or had more land, but it was the way they managed their finances that set them apart from others.

During the normal year, Muslims and Christians relied more on money lenders, whereas Hindus primarily obtained loans from commercial banks (Table 8.1). But, during the bad year, when crops failed and debts piled up, more people than ever went to money lenders to borrow money (Table 8.2). And, among them, Hindus had the most debts. This was because they were using the borrowed money to pay off the previous loan they had taken from formal institutions. But what was even more concerning was the increase in informal sources of credit during these tough times. Money lenders, who charged very high interest rates, became the only option for many farmers. And it was not just the Hindus, but all religious groups were heavily indebted to these money lenders.

TABLE 8.1
Indebtedness (Rs) of Households by Religion in Normal Year (2020–2021)

	Religion							
	Hindu		Muslim		Christian		Total	
Source of Credit	Mean	%	Mean	%	Mean	%	Mean	%
Commercial banks (loan taken)	1,98,731	37	1,43,750	23	1,02,000	52	1,95,270	37
Commercial banks (outstanding)	2,04,234	25	1,50,909	21	97,429	30	2,00,507	25
Regional rural bank (loan taken)	1,54,335	16	79,000	10	90,000	9	1,52,027	15
Regional rural bank (Outstanding)	1,51,450	11	79,000	10	1,00,000	4	1,48,781	11
Co-operative societies/banks (loan taken)	1,34,399	13	1,02,364	21	1,15,000	9	1,32,296	13
Co-operative societies/banks (outstanding)	1,32,965	8	1,17,800	19	1,17,500	9	1,31,516	9
SHGs (loan taken)	66,662	6	1,20,000	4	50,000	9	67,603	5
SHGs outstanding)	67,676	3	1,40,000	2		0	69,743	2
Money lender (loan taken)	2,57,409	16	1,76,250	15	2,88,000	22	2,55,279	16
Money lender (outstanding)	2,78,644	12	2,13,333	12	3,81,667	13	2,78,163	12
Relatives/friends (loan taken)	2,35,854	6	1,30,000	10	1,00,000	4	2,28,295	6
Relatives/friends (outstanding)	2,61,809	3	1,33,333	6	1,00,000	4	2,51,078	4
Traders (loan taken)	2,86,897	2		0		0	2,86,897	2
Traders (outstanding)	3,81,429	1		0		0	3,81,429	0.2
Others (loan taken)	2,72,897	2		0		0	2,72,897	2
Others (outstanding)	1,99,909	1		0		0	1,99,909	1
Total loan taken (formal)	1,90,315	63	1,39,440	48	1,23,857	61	1,87,821	62
Total outstanding (formal)	1,88,407	43	1,40,542	46	1,13,000	39	1,85,431	43
Total loan taken (informal)	2,69,018	25	1,71,667	23	3,08,000	22	2,66,306	25
Total outstanding (informal)	2,76,960	17	1,86,667	17	3,11,250	17	2,74,134	17

Note: % indicates the proportion of households within each category.

TABLE 8.2
Indebtedness (Rs.) by Religion in Bad Year (2021–2022)

	Religion							
	Hindu		Muslim		Christian		Total	
Source of Credit	**Mean**	**%**	**Mean**	**%**	**Mean**	**%**	**Mean**	**%**
Commercial banks (loan taken)	2,05,058	23	1,11,571	13	80,000	26	2,00,682	23
Commercial banks (outstanding)	2,09,856	31	1,16,333	23	1,02,000	30	2,05,491	30
Regional rural bank (loan taken)	1,65,224	11	86,333	12	80,000	4	1,61,597	11
Regional rural bank (outstanding)	1,65,606	13	91,600	10	90,000	9	1,62,742	13
Co-operative societies/banks (loan taken)	1,66,180	8	31,400	10	2,00,000	4	1,60,709	8
Co-operative societies/banks (outstanding)	1,79,431	9	27,400	10	2,00,000	4	1,73,698	9
SHGs (loan taken)	76,130	4	66,667	6	50,000	13	74,350	4
SHGs (outstanding)	62,345	2	35,000	4	15,000	9	57,818	2
Money lender (loan taken)	3,20,149	23	2,00,833	23	1,80,000	22	3,13,745	23
Money lender (outstanding)	3,21,570	25	2,17,143	27	3,16,667	26	3,17,450	25
Relatives/friends (loan taken)	2,12,768	4	2,50,000	6		0	2,14,661	4
Relatives/friends (outstanding)	2,39,177	5	3,16,667	6	1,00,000	4	2,40,591	5
Traders (loan taken)	1,29,565	2		0		0	1,29,565	2
Traders (outstanding)	2,50,556	1		0		0	2,50,556	1
Others (loan taken)	2,58,346	2		0		0	2,58,346	2
Others (outstanding)	2,70,128	3		0		0	2,70,128	3
Total loan taken (formal)	2,02,958	40	97,412	33	1,13,750	35	1,98,534	40
Total outstanding (formal)	2,13,032	49	1,03,050	38	1,40,500	35	2,08,950	48
Total loan taken (informal)	2,99,461	30	2,10,667	29	1,80,000	22	2,94,974	30
Total outstanding (informal)	3,12,167	33	2,34,706	33	3,33,333	26	3,09,641	33

Note: % indicates the proportion of households within each category.

Social Groups

As our study investigated the borrowing habits of farmers from different social groups, we discovered that even though banks were lending to farmers, it was not reaching everyone. This means that caste still affects who can get loans. As a result, large proportions of farmers from socially marginalized castes (SC and ST) were outside the fold of formal credit (Karthick and Madheswaran, 2018). We saw that, regardless of their caste, farmers mostly preferred getting loans from formal sources like commercial banks, but money lenders often gave higher amounts to all the farmers (Table 8.3). Co-operative societies also provided loans, but they discriminated against lower-caste farmers in lending (Jodhka, 1995; Kumar, 2013). These differences in loan access across caste groups might be because of the differences in crop yields and land productivity between them (Rao, 2017). However, in 2022, no matter their caste, fewer farmers used both commercial banks and co-operative societies and more turned to money lenders (Table 8.4). This trend was consistent across different caste categories.

TABLE 8.3
Indebtedness (Rs.) by Social Group in Normal Year (2020–2021)

	Caste							
	SC		ST		OBC		General	
Source of Credit	**Mean**	**%**	**Mean**	**%**	**Mean**	**%**	**Mean**	**%**
Commercial banks (loan taken)	1,09,070	33	1,45,658	26	2,04,864	35	2,10,240	42
Commercial banks (outstanding)	94,375	18	1,53,250	21	2,21,531	25	2,06,460	28
Regional rural bank (loan taken)	1,12,750	15	1,25,238	14	1,54,815	14	1,63,598	17
Regional rural bank (outstanding)	1,20,455	8	1,30,647	12	1,55,103	10	1,52,449	12
Co-operative societies/banks (loan taken)	55,000	6	78,684	13	1,47,084	15	1,37,684	13
Co-operative societies/banks (outstanding)	65,833	5	89,167	8	1,54,750	11	1,21,271	8
SHGs (loan taken)	93,167	9	53,333	8	66,719	6	62,727	4
SHGs (outstanding)	96,833	5	40,000	3	64,118	3	81,429	1
Money lender (loan taken)	2,17,321	21	2,18,958	16	2,41,042	17	2,94,118	15
Money lender (outstanding)	2,40,000	15	2,46,053	13	2,52,056	13	3,31,492	10
Relatives/friends (loan taken)	1,13,000	8	2,56,364	8	2,31,111	6	2,52,258	5
Relatives/friends (outstanding)	1,16,667	2	2,77,000	7	2,68,696	4	2,33,667	3
Traders (loan taken)	1,16,667	2		0	2,77,059	3	3,62,222	2
Traders (outstanding)		0		0	5,60,000	1	1,43,333	1
Others (loan taken)	3,00,000	1	90,000	1	2,52,000	2	2,97,000	3
Others (outstanding)	3,00,000	1		0	1,05,500	1	2,46,167	1
Total loan taken (formal)	1,18,097	55	1,28,750	55	1,95,408	62	2,06,317	66
Total outstanding (formal)	1,03,773	34	1,34,808	41	2,00,196	45	1,96,667	44
Total loan taken (informal)	2,01,667	30	2,26,806	25	2,59,613	27	3,03,224	23
Total outstanding (informal)	2,27,083	18	2,56,724	20	2,64,608	18	3,05,155	14

Note: % indicates the proportion of households within each category.

TABLE 8.4
Indebtedness by Social Group in Bad Year (2021–2022)

	Social Category							
	SC		ST		OBC		General	
Source of Credit	**Mean**	**%**	**Mean**	**%**	**Mean**	**%**	**Mean**	**%**
Commercial banks (loan taken)	1,46,192	20	1,33,226	21	2,19,991	20	2,09,464	26
Commercial banks (outstanding)	1,32,778	27	1,11,806	25	2,25,087	27	2,20,159	36
Regional rural bank (loan taken)	1,17,143	11	1,50,000	11	1,50,066	11	1,85,587	11
Regional rural bank (outstanding)	1,32,632	15	1,40,000	14	1,50,881	12	1,87,333	13
Co-operative societies/banks (loan taken)	77,571	5	74,063	11	1,70,000	8	1,93,298	8
Co-operative societies/banks (outstanding)	81,667	5	86,154	9	1,94,345	10	1,83,173	9
SHGs (loan taken)	53,571	5	71,429	5	74,625	4	81,591	4
SHGs outstanding)	27,000	4	92,500	3	47,714	2	73,500	2
Money lender (loan taken)	2,25,000	26	2,11,750	27	3,77,592	26	2,93,313	19
Money lender (outstanding)	2,46,622	28	1,98,068	30	3,85,742	29	2,89,839	20

(*Continued*)

TABLE 8.4 (*Continued*)
Indebtedness by Social Group in Bad Year (2021–2022)

	Social Category							
	SC		ST		OBC		General	
Source of Credit	**Mean**	**%**	**Mean**	**%**	**Mean**	**%**	**Mean**	**%**
Relatives/friends (loan taken)	1,05,000	5	2,32,222	6	2,69,000	4	1,69,474	3
Relatives/friends (outstanding)	1,05,000	5	2,47,083	8	2,65,143	5	2,43,000	3
Traders (loan taken)	75,000	3		0	1,65,833	2	98,571	1
Traders (outstanding)	66,667	2		0	1,83,333	1	3,56,667	2
Others (loan taken)		0	90,000	1	1,90,625	1	3,00,118	3
Others (outstanding)	3,00,000	2	90,000	1	2,61,071	2	2,81,364	4
Total loan taken (formal)	1,32,479	37	1,32,500	42	2,04,645	38	2,23,049	42
Total outstanding (formal)	1,36,638	44	1,30,077	45	2,13,023	47	2,36,834	51
Total loan taken (informal)	1,99,535	33	2,13,000	34	3,51,583	33	2,79,087	26
Total outstanding (informal)	2,29,457	35	2,06,491	39	3,69,729	36	2,93,703	28

Note: % indicates the proportion of households within each category.

Landholding Groups (Own Land)

The stories of the farmers owing their land among various landholding groups disheartened us heavily. We noticed that farmers with large landholding groups were more heavily in debt (Singh, 2006; Singh et al., 2014) and had outstanding loans (Rao and Suri, 2006). In the year 2020–2021, almost half of the farmers, from all types of landholdings, took loans from commercial banks, with large farmers borrowing the most around Rs. 5 lakh (Table 8.5). Surprisingly, 42% of these borrowers were medium landholders, showing their heavy reliance on loans for livelihood. Although commercial banks were the main source of loans for all farmers, money

TABLE 8.5
Indebtedness (Rs.) by Land Size Category in Normal Year (2020–2021)

	Land Size									
	Marginal		Small		Semi-Medium		Medium		Large	
Source of Credit	**Mean**	**%**	**Mean**	**%**	**Mean**	**%**	**Mean**	**%**	**Mean**	**%**
Commercial banks (loan taken)	1,14,091	28	1,17,241	31	1,41,906	37	2,23,749	42	4,92,960	50
Commercial banks (outstanding)	1,19,595	19	1,26,711	22	1,44,761	25	2,26,675	29	5,34,677	31
Regional rural bank (loan taken)	77,368	10	78,486	15	1,20,190	17	2,06,449	17	3,45,333	15
Regional rural bank (outstanding)	85,000	9	80,667	12	1,21,737	12	2,09,425	10	3,52,727	11
Co-operative societies/ banks (loan taken)	71,773	11	1,05,152	13	1,20,885	13	1,42,255	13	2,90,667	15

(*Continued*)

TABLE 8.5 (*Continued*)
Indebtedness (Rs.) by Land Size Category in Normal Year (2020–2021)

	Land Size									
	Marginal		Small		Semi-Medium		Medium		Large	
Source of Credit	**Mean**	%	**Mean**	%	**Mean**	%	**Mean**	%	**Mean**	%
Co-operative societies/ banks (outstanding)	69,667	8	85,542	9	1,28,050	9	1,36,135	9	3,31,400	10
SHGs (loan taken)	80,571	7	70,667	6	56,400	5	66,579	5	82,000	5
SHGs outstanding)	67,625	4	1,00,000	2	65,500	2	63,182	3	75,000	2
Money lender (loan taken)	1,92,500	21	1,78,438	19	2,42,083	16	3,52,727	13	3,75,000	16
Money lender (outstanding)	1,82,179	14	1,92,051	15	2,58,302	12	4,06,076	10	4,43,333	12
Relatives/friends (loan taken)	1,69,286	7	1,73,958	9	2,44,423	6	2,36,111	4	4,90,000	6
Relatives/friends (outstanding)	1,94,444	5	2,03,333	6	2,74,333	3	2,72,222	2	4,80,000	3
Traders (loan taken)	1,06,667	2	2,62,500	3	2,76,667	2	1,30,000	2	12,50,000	2
Traders (outstanding)	2,00,000	1	30,000	0	2,70,000	0	2,00,000	0	15,00,000	1
Others (loan taken)	1,10,000	1	1,16,000	2	1,26,667	2	4,53,077	3	2,00,000	1
Others (outstanding)	1,37,500	2	2,11,000	1	1,00,000	0	3,42,333	1		0
Total loan taken (formal)	1,07,340	49	1,18,411	55	1,42,768	65	2,24,799	67	4,73,945	72
Total outstanding (formal)	1,05,014	36	1,15,195	41	1,45,370	44	2,19,734	46	4,78,380	50
Total loan taken (informal)	1,92,768	28	1,88,938	32	2,42,611	25	3,62,500	21	4,65,600	25
Total outstanding (informal)	1,85,390	21	1,92,842	22	2,57,569	16	3,79,623	13	5,16,250	16

Note: % indicates the proportion of households within each category.

lenders were often the last resort, especially for marginal and small farmers, because this informal way of borrowing money was crucial for them when they needed to get loans (Chhikara and Kodan, 2013). But the situation worsened in the following year. Despite fewer farmers using commercial bank services, the amount of loans taken increased for marginal, small, and medium farmers (Table 8.6) mainly due to the huge crop losses in the previous years. As a result, farmers had to resort to money lenders as their go-to option in need of quick cash to repay previous loans.

Landholding Groups (Total Cultivated Land)

Similarly, in this group also, small and marginal farmers mostly depended on informal sources like relatives, commercial banks, and sometimes money lenders for loans. Larger landholders also sought loans from money lenders for a quite higher amount, although they preferred formal sources (Table 8.7). However, in 2021, the trend shifted, with both large and medium farmers turning to informal sources for loans underscoring the need for financial support (Table 8.8). To investigate this incidence, we noted that indebtedness was the root cause of the vulnerability of smallholders especially in dryland areas (Ramprasad, 2019).

TABLE 8.6
Indebtedness (Rs.) by Own-Land Size Category in Bad Year (2021–2022)

	Land Size									
	Marginal		Small		Semi-Medium		Medium		Large	
Source of Credit	Mean	%	Mean	%	Mean	%	Mean	%	Mean	%
Commercial banks (loan taken)	1,03,973	19	2,75,373	20	1,47,710	23	2,03,730	25	3,98,077	26
Commercial banks (outstanding)	1,19,862	29	2,39,725	27	1,46,932	29	2,13,409	32	4,43,667	39
Regional rural bank (loan taken)	92,214	7	90,536	11	1,49,667	13	1,98,049	10	3,60,000	11
Regional rural bank (outstanding)	96,067	8	90,441	13	1,44,056	16	2,08,021	12	3,49,286	14
Co-operative societies/ banks (loan taken)	1,95,250	6	81,579	7	1,50,028	8	1,64,100	10	2,94,500	10
Co-operative societies/ banks (outstanding)	1,68,357	7	91,762	8	1,68,119	9	1,52,659	10	4,36,636	11
SHGs (loan taken)	58,417	6	65,000	6	67,647	4	1,06,250	3	90,000	4
SHGs outstanding)	52,143	4	42,889	4	63,143	2	59,444	2	1,80,000	1
Money lender (loan taken)	2,95,556	27	1,73,922	20	2,92,358	23	3,36,870	25	6,80,455	22
Money lender (outstanding)	3,02,232	28	1,64,455	22	2,93,947	25	3,52,518	27	6,17,037	27
Relatives/friends (loan taken)	1,09,286	4	2,06,154	5	2,35,000	4	2,26,000	4	2,66,667	6
Relatives/friends (outstanding)	1,25,692	7	2,35,714	5	2,76,053	4	2,33,333	4	4,40,000	5
Traders (loan taken)	66,667	2	1,05,000	2	1,71,429	2	1,38,750	2	50,000	1
Traders (outstanding)	66,667	2	2,87,500	2	3,37,500	1	1,52,000	1	5,25,000	2
Others (loan taken)		0	1,60,500	2	1,42,500	2	3,52,917	3	3,50,000	2
Others (outstanding)	1,21,667	3	2,14,000	2	1,95,000	3	4,13,929	3	3,00,000	2
Total loan taken (formal)	1,29,873	32	1,99,000	38	1,61,526	42	2,14,982	41	3,91,444	45
Total outstanding (formal)	1,37,222	41	1,90,687	45	1,65,004	49	2,23,627	49	4,60,610	58
Total loan taken (informal)	2,72,984	31	1,75,167	28	2,76,176	30	3,26,323	32	5,77,333	30
Total outstanding (informal)	2,66,973	37	1,89,156	30	2,86,791	32	3,51,309	34	6,03,235	34

Note: % indicates the proportion of households within each category.

Poverty Groups

Moving forward to the poverty groups, we discovered that poor farmers faced more challenges accessing banks for loans compared to non-poor farmers (Tran et al., 2016). We felt disheartened to see that most of the poor farmers had to rely on

TABLE 8.7
Indebtedness (Rs.) by Land Cultivated Size Group in Normal Year (2020–2021)

	Land Size									
	Marginal		Small		Semi-Medium		Medium		Large	
Source of Credit	**Mean**	%	**Mean**	%	**Mean**	%	**Mean**	%	**Mean**	%
Commercial banks (loan taken)	1,07,083	30	1,04,125	25	1,25,128	35	1,93,763	42	3,58,990	44
Commercial banks (outstanding)	92,813	20	1,12,938	21	1,31,621	23	1,93,908	28	3,92,939	28
Regional rural bank (loan taken)	72,222	11	72,414	15	1,15,395	19	1,85,284	13	2,46,316	16
Regional rural bank (outstanding)	86,111	11	73,300	10	1,14,089	14	1,78,091	9	2,53,654	11
Co-operative societies/ banks (loan taken)	69,929	18	93,962	13	84,477	11	1,42,217	14	2,31,970	14
Co-operative societies/ banks (outstanding)	63,889	11	72,864	11	75,357	7	1,41,070	8	2,59,042	10
SHGs (loan taken)	94,000	9	71,667	3	56,731	6	54,615	5	99,231	6
SHGs outstanding)	67,750	5	1,00,000	2	61,250	3	55,000	2	1,06,000	2
Money lender (loan taken)	1,51,818	14	1,43,710	16	2,00,878	18	2,83,889	14	4,01,111	19
Money lender (outstanding)	1,62,222	11	1,37,500	13	2,27,018	14	3,21,597	10	4,69,500	12
Relatives/friends (loan taken)	95,000	8	1,71,250	10	2,17,045	5	2,47,500	6	3,65,833	5
Relatives/friends (outstanding)	1,12,500	5	1,82,143	7	2,51,154	3	3,03,571	3	3,81,667	3
Traders (loan taken)	1,00,000	1	75,000	2	2,07,143	2	1,55,833	2	9,20,000	2
Traders (outstanding)		0	30,000	1	5,00,000	0	1,60,000	1	15,00,000	0
Others (loan taken)		0	77,000	1	1,42,500	1	2,58,889	4	5,06,000	2
Others (outstanding)	50,000	1	22,000	1	1,00,000	0	2,75,400	1	2,75,000	1
Total loan taken (formal)	1,07,044	56	1,00,717	51	1,22,027	64	1,98,557	64	3,54,335	67
Total outstanding (formal)	91,353	43	99,829	41	1,22,814	43	1,93,741	43	3,63,667	47
Total loan taken (informal)	1,37,647	21	1,48,821	29	2,02,430	26	2,89,744	23	4,73,710	27
Total outstanding (informal)	1,40,000	18	1,47,071	22	2,31,575	18	3,10,676	15	4,85,722	16

Note: % indicates the proportion of households within each category.

informal sources for loans due to limited access to formal banking services. Several poor farmers shared their stories with us. Many of them did not own any land or property, making it difficult for them to borrow more from commercial banks. This resulted in higher debts for poor farmers. The pressure on poor households appeared to worsen as the bad year progressed; they turned to unofficial sources like money lenders and dealers for loans.

The real shocker came when we looked at the amount of loans taken by poor and non-poor farmers from commercial banks. In 2020–2021, the average amount of loans taken by poor farmers was around Rs. 2.07 lakh, while non-poor farmers took an average of Rs. 1.8 lakh (Table 8.9). It meant that poor farmers were taking on more debt, despite having limited access to banking services. The trend continued in 2021–2022, with poor farmers struggling to break free from the cycle of debt, while non-poor farmers were comparatively better off (Tables 8.10–8.12).

Tenancy Groups

Somehow, some of us had assumed that all the farmers faced similar challenges when it came to borrowing money. But our research proved otherwise. We found a distinct pattern between the two. The owner-farmers could easily get loans from formal sources like commercial banks and co-operative organizations, but tenant farmers had to rely on informal channels (Table 8.13). They struggled to get proper banking services, but, since they did not own land, they were not eligible for crop loans

TABLE 8.8
Indebtedness (Rs.) by Cultivated Land Size Category in Bad Year (2021–2022)

	Land Size									
	Marginal		Small		Semi-Medium		Medium		Large	
Source of Credit	Mean	%	Mean	%	Mean	%	Mean	%	Mean	%
Commercial banks (loan taken)	98,529	21	3,36,294	17	1,27,901	20	1,83,992	26	2,88,661	25
Commercial banks (outstanding)	93,269	33	2,85,933	23	1,31,020	25	1,91,235	34	3,12,586	38
Regional rural bank (loan taken)	1,02,750	5	91,429	11	1,40,463	13	1,73,830	9	2,42,857	12
Regional rural bank (Outstanding)	1,02,200	6	90,800	13	1,36,141	16	1,83,396	10	2,40,143	15
Co-operative societies/banks (loan taken)	2,95,000	5	73,615	7	1,23,429	7	1,71,731	10	2,14,000	9
Co-operative societies/banks (outstanding)	2,16,167	8	72,938	8	1,24,200	7	1,78,377	10	2,81,792	10
SHGs (loan taken)	98,333	4	60,000	3	58,438	4	78,185	5	91,111	4
SHGs outstanding)	1,22,500	3	53,333	2	56,875	2	42,375	3	92,500	2
Money lender (loan taken)	1,89,286	18	1,35,000	16	2,10,298	21	2,80,969	26	5,91,370	31
Money lender (outstanding)	2,03,750	15	1,29,706	17	2,24,202	23	2,85,964	27	5,62,471	37
Relatives/friends (loan taken)	1,31,250	5	1,74,545	6	1,52,308	3	2,52,000	4	2,90,909	5
Relatives/friends (outstanding)	82,333	8	2,10,769	7	2,19,667	4	2,35,000	4	3,87,500	5
Traders (loan taken)		0	73,333	2	1,20,000	2	1,59,091	2	50,000	0.2
Traders (outstanding)		0	50,000	2	1,50,000	1	1,43,333	1	6,87,500	2
Others (loan taken)		0	2,06,000	1	1,10,000	1	2,62,188	3	3,56,000	2
Others (outstanding)	50,000	1	2,80,000	1	1,52,000	1	2,34,773	4	4,28,000	4
Total loan taken (formal)	1,42,440	31	2,24,785	33	1,39,600	39	1,97,844	43	2,92,232	43
Total outstanding (formal)	1,31,706	43	2,05,800	41	1,42,707	45	2,09,448	49	3,21,301	57
Total loan taken (informal)	1,76,389	23	1,43,340	24	1,97,500	26	2,81,129	34	5,41,573	38
Total outstanding (informal)	1,57,316	24	1,48,627	26	2,19,322	29	2,80,348	35	5,66,571	45

Note: % indicates the proportion of households within each category.

TABLE 8.9
Indebtedness (Rs.) by Poverty Status in Normal Year

	Poverty status, 2020–2021			
	Non-poor		Poor	
Source of Credit	**Mean**	**%**	**Mean**	**%**
Commercial banks (loan taken)	2,00,815	38	1,83,556	36
Commercial banks (outstanding)	2,03,613	25	1,94,346	25
Regional rural bank (loan taken)	1,48,825	16	1,59,615	14
Regional rural bank (outstanding)	1,50,353	12	1,44,103	8
Co-operative societies/banks (loan taken)	1,32,115	12	1,32,575	15
Co-operative societies/banks (outstanding)	1,28,224	8	1,36,520	11
SHGs (loan taken)	62,759	6	81,650	4
SHGs (outstanding)	66,600	3	77,600	2
Money lender (loan taken)	2,64,487	16	2,36,623	16
Money lender (outstanding)	2,90,082	12	2,57,016	13
Relatives/friends (loan taken)	2,17,404	5	2,44,028	8
Relatives/friends (outstanding)	2,51,094	3	2,51,053	4
Traders (loan taken)	3,18,182	2	1,88,571	1
Traders (outstanding)	5,25,000	0	1,90,000	1
Others (loan taken)	2,18,000	2	3,31,714	3
Others (outstanding)	2,75,000	1	1,09,800	1
Total loan taken (formal)	1,87,693	64	1,88,096	59
Total outstanding (formal)	1,83,637	44	1,89,345	41
Total loan taken (informal)	2,68,820	25	2,61,659	27
Total outstanding (informal)	2,91,294	16	2,45,213	19

Notes: Household categorized as poor and non-poor based on income in year 2020–2021. % indicates the proportion of households within each category.

TABLE 8.10
Indebtedness (Rs.) in Bad Year (2021–2022) for Households Who Are Poor in 2020–2021

	Poverty status, 2020–2021			
	Non-poor		Poor	
Source of Credit	**Mean**	**%**	**Mean**	**%**
Commercial banks (loan taken)	2,13,363	24	1,71,827	23
Commercial banks (outstanding)	2,24,208	31	1,63,235	31
Regional rural bank (loan taken)	1,64,774	12	1,52,231	11
Regional rural bank (outstanding)	1,66,208	14	1,54,077	13
Co-operative societies/banks (loan taken)	1,61,721	9	1,57,903	8
Co-operative societies/banks (outstanding)	1,57,389	10	2,19,265	9

(Continued)

TABLE 8.10 (*Continued*)
Indebtedness (Rs.) in Bad Year (2021–2022) for Households Who Are Poor in 2020–2021

	Poverty status, 2020–2021			
	Non-poor		Poor	
Source of Credit	Mean	%	Mean	%
SHGs (loan taken)	75,750	5	68,750	4
SHGs outstanding)	63,320	3	40,625	2
Money lender (loan taken)	3,29,708	26	2,65,663	23
Money lender (outstanding)	3,32,230	28	2,74,076	25
Relatives/friends (loan taken)	2,25,395	4	1,95,238	4
Relatives/friends (outstanding)	2,45,951	4	2,31,800	5
Traders (loan taken)	1,41,053	2	75,000	2
Traders (outstanding)	2,58,667	2	2,10,000	1
Others (loan taken)	2,37,667	2	2,86,545	2
Others (outstanding)	2,13,654	3	3,83,077	3
Total loan taken (formal)	2,05,056	43	1,81,503	40
Total outstanding (formal)	2,15,074	51	1,93,466	49
Total loan taken (informal)	3,09,672	33	2,55,190	30
Total outstanding (informal)	3,19,360	36	2,83,876	33

Notes: Household categorized as poor and non-poor based on income in year 2020–2021. % indicates the proportion of households within each category.

TABLE 8.11
Indebtedness (Rs.) in Normal Year (2020–2021) for Households Who are Poor in 2021–2022

	Poverty status, 2021–2022			
	Non-poor		Poor	
Source of Credit	Mean	%	Mean	%
Commercial banks (loan taken)	2,07,116	35	1,83,058	39
Commercial banks (outstanding)	2,09,122	23	1,92,540	28
Regional rural bank (loan taken)	1,46,363	16	1,59,421	14
Regional rural bank (outstanding)	1,48,851	12	1,48,691	10
Co-operative societies/banks (loan taken)	1,24,495	13	1,41,172	13
Co-operative societies/banks (outstanding)	1,21,058	9	1,44,175	9
SHGs (loan taken)	65,098	7	72,333	4
SHGs (outstanding)	71,316	3	67,875	2
Money lender (loan taken)	2,51,653	16	2,59,196	17
Money lender (outstanding)	2,60,767	12	2,97,256	12

(*Continued*)

TABLE 8.11 (*Continued*)
Indebtedness (Rs.) in Normal Year (2020–2021) for Households Who are Poor in 2021–2022

	Poverty status, 2021–2022			
	Non-poor		Poor	
Source of Credit	**Mean**	**%**	**Mean**	**%**
Relatives/friends (loan taken)	2,38,143	5	2,21,792	8
Relatives/friends (outstanding)	2,78,913	3	2,28,214	4
Traders (loan taken)	3,83,750	2	1,67,692	2
Traders (outstanding)	5,60,000	1	1,43,333	0
Others (loan taken)	2,62,727	1	2,79,111	3
Others (outstanding)	2,37,500	1	1,78,429	1
Total loan taken (formal)	1,88,359	63	1,87,200	62
Total outstanding (formal)	1,83,209	42	1,87,849	44
Total loan taken (informal)	2,70,540	23	2,62,235	28
Total outstanding (informal)	2,77,933	16	2,70,367	18

Notes: Household categorized as poor and non-poor based on income in year 2021–2022. % indicates the proportion of households within each category.

TABLE 8.12
Indebtedness (Rs.) by Poverty Level in Bad Year (2021–2022) for Households Who Are Poor in 2021–2022

	Poverty status, 2021–2022			
	Non-poor		Poor	
Source of Credit	**Mean**	**%**	**Mean**	**%**
Commercial Banks (loan taken)	2,25,337	24	1,69,993	22
Commercial Banks (outstanding)	2,35,112	33	1,64,350	27
Regional Rural Bank (loan taken)	1,52,783	12	1,74,677	9
Regional Rural Bank (Outstanding)	1,54,769	14	1,73,372	12
Cooperative societies/banks (loan taken)	1,56,310	9	1,67,500	7
Cooperative societies/banks (outstanding)	1,41,797	10	2,24,100	8
SHGs (loan taken)	79,795	6	59,375	2
SHGs (outstanding)	61,217	3	50,000	2
Money lender (loan taken)	3,10,813	26	3,17,786	21
Money lender (outstanding)	3,04,305	28	3,36,242	22
Relatives/ Friends (loan taken)	2,34,783	3	2,01,806	5
Relatives/ Friends (outstanding)	2,48,281	4	2,33,353	5
Traders (loan taken)	1,18,333	2	1,41,818	2

(*Continued*)

TABLE 8.12 (*Continued*)
Indebtedness (Rs.) by Poverty Level in Bad Year (2021–2022) for Households Who Are Poor in 2021–2022

	Poverty status, 2021–2022			
	Non-poor		Poor	
Source of Credit	**Mean**	**%**	**Mean**	**%**
Traders (outstanding)	3,06,154	2	1,06,000	1
Others (loan taken)	2,65,000	2	2,51,692	2
Others(outstanding)	2,20,833	2	3,12,381	3
Total loan taken(formal)	2,10,321	43	1,82,475	36
Total outstanding(formal)	2,18,165	53	1,95,754	42
Total loan taken (informal)	2,96,422	31	2,93,197	29
Total outstanding (informal)	3,01,183	35	3,20,808	31

Notes: Households categorised as poor and non-poor based on income in year 2021–2022.
% indicates the proportion of households within each category.

TABLE 8.13
Indebtedness (Rs.) by Tenancy Status in Normal Year (2020–2021)

	Owner-cultivator		Pure tenant-cultivator	
Source of Credit	**Mean**	**%**	**Mean**	**%**
Commercial banks (loan taken)	1,97,671	38	1,18,750	22
Commercial banks (outstanding)	2,02,699	26	1,31,364	15
Regional rural bank (loan taken)	1,52,599	16	90,000	3
Regional rural bank (outstanding)	1,49,549	11	90,000	3
Co-operative societies/banks (loan taken)	1,33,761	13	88,333	8
Co-operative societies/banks (outstanding)	1,32,854	9	76,667	4
SHGs (loan taken)	68,419	5	52,500	6
SHGs (outstanding)	71,848	2	35,000	3
Money lender (loan taken)	2,61,033	16	1,94,000	28
Money lender (outstanding)	2,88,597	12	1,25,455	15
Relatives/friends (loan taken)	2,28,795	6	2,20,000	7
Relatives/friends (outstanding)	2,48,021	4	3,00,000	4
Traders (loan taken)	2,96,429	2	20,000	1
Traders (outstanding)	3,81,429	1		0
Others (loan taken)	2,84,963	2	1,10,000	3
Others (outstanding)	1,99,909	1		0
Total loan taken (formal)	1,90,451	64	1,04,444	38
Total outstanding (formal)	1,87,487	44	1,13,235	24
Total loan taken (informal)	2,70,611	25	2,08,800	35
Total outstanding (informal)	2,81,058	17	1,62,857	19

Notes: Pure tenants are with no owned land.
% indicates the proportion of households within each category.

TABLE 8.14
Indebtedness (Rs.) by Tenancy Status in Bad Year (2021–2022)

	Owner-cultivator		Pure tenant-cultivator	
Source of Credit	**Mean**	%	**Mean**	%
Commercial banks (loan taken)	2,03,945	23	1,16,667	17
Commercial banks (outstanding)	2,08,097	31	1,45,833	25
Regional rural bank (loan taken)	1,61,771	11	1,35,000	1
Regional rural bank (outstanding)	1,63,578	13	87,500	3
Co-operative societies/banks (loan taken)	1,62,973	8	1,10,000	7
Co-operative societies/banks (outstanding)	1,75,696	9	1,11,250	6
SHGs (loan taken)	77,364	4	41,200	7
SHGs (outstanding)	59,290	2	35,000	3
Money lender (loan taken)	3,07,859	23	3,77,857	39
Money lender (outstanding)	3,10,922	25	3,92,414	40
Relatives/friends (loan taken)	2,18,684	4	1,00,000	3
Relatives/friends (outstanding)	2,46,435	5	1,50,000	6
Traders (loan taken)	1,39,000	1	66,667	4
Traders (outstanding)	2,87,333	1	66,667	4
Others (loan taken)	2,58,346	2		0
Others (outstanding)	2,79,865	3	90,000	3
Total loan taken (formal)	2,01,605	41	1,14,550	28
Total outstanding (formal)	2,11,216	49	1,44,130	32
Total loan taken (informal)	2,90,373	30	3,54,194	43
Total outstanding (informal)	3,05,449	32	3,63,529	47

Note: % indicates the proportion of households within each category.

from banks (Sreenivasulu, 2023). This tenancy farming was a burden for farmers. The high rental rate, coupled with the dependency on informal sources of credit, made life miserable for them, and sadly it even led some farmers to take drastic measures like suicide (Arora, 2021; Reddy et al., 2020; Chand et al., 2017). When we spoke to the owner-farmers who rented their land to these tenants, they admitted that they knew about the tenants' problems but were helpless because they needed the rent money to sustain themselves. It was a difficult situation for both groups (Table 8.14).

Irrigation Groups

More dryland farmers took loans mostly from commercial banks compared to those with irrigated land in 2020–2021 (Table 8.15). Even the indebtedness was higher in dryland farmers, which was also discovered by Babu (2018). However, in the following year, there was a drastic change; a noticeable decrease was found among dryland and irrigated farmers who were using bank loans, but irrigated farmers continued to borrow higher amounts (Table 8.16). Upon further investigation, we found that this trend was a result of the increasing debt among both groups of farmers, leading many to struggle with repayment and turn to informal income sources, especially money lenders.

TABLE 8.15
Indebtedness (Rs.) by Dry and Irrigated Farmers in Normal Year (2020–2021)

	Dryland		Irrigated Land	
Source of Credit	**Mean**	%	**Mean**	%
Commercial banks (loan taken)	1,95,815	39	1,94,905	36
Commercial banks (outstanding)	2,05,268	28	1,97,034	24
Regional rural bank (loan taken)	1,37,744	15	1,60,577	16
Regional rural bank (outstanding)	1,36,596	9	1,54,083	12
Co-operative societies/banks (loan taken)	1,20,645	14	1,40,345	13
Co-operative societies/banks (outstanding)	1,14,558	10	1,43,432	8
SHGs (loan taken)	61,857	6	72,279	5
SHGs (outstanding)	67,273	2	70,875	3
Money lender (loan taken)	2,03,500	17	2,87,867	16
Money lender (outstanding)	2,07,136	11	3,15,248	13
Relatives/friends (loan taken)	2,15,741	5	2,33,852	7
Relatives/friends (outstanding)	2,15,000	1	2,56,818	5
Traders (loan taken)	1,61,429	1	3,26,818	3
Traders (outstanding)	2,35,000	1	5,76,667	0.2
Others (loan taken)	2,14,800	3	3,35,143	2
Others (outstanding)	1,42,714	1	3,00,000	0.2
Total loan taken (formal)	1,85,773	63	1,89,128	62
Total outstanding (formal)	1,91,793	42	1,81,620	44
Total loan taken (informal)	2,21,031	24	2,91,394	26
Total outstanding (informal)	2,06,118	14	3,05,847	19

Note: % indicates the proportion of households within each category.

TABLE 8.16
Indebtedness (Rs.) for Dry and Irrigated Farmers in Bad Year (2021–2022)

	Dryland		Irrigated Land	
Source of Credit	**Mean**	%	**Mean**	%
Commercial banks (loan taken)	1,77,370	23	2,15,943	22
Commercial banks (outstanding)	1,89,591	31	2,15,988	30
Regional rural bank (loan taken)	1,67,582	10	1,58,273	11
Regional rural bank (outstanding)	1,66,121	11	1,61,161	14
Co-operative societies/banks (loan taken)	1,37,185	10	1,80,873	7
Co-operative societies/banks (outstanding)	1,45,450	11	1,98,261	8
SHGs (loan taken)	60,026	7	1,00,952	2
SHGs (outstanding)	42,611	3	76,067	2
Money lender (loan taken)	3,03,636	24	3,20,383	23

(Continued)

TABLE 8.16 (*Continued*)
Indebtedness (Rs.) for Dry and Irrigated Farmers in Bad Year (2021–2022)

	Dryland		Irrigated Land	
Source of Credit	**Mean**	**%**	**Mean**	**%**
Money lender (outstanding)	2,91,014	25	3,33,737	26
Relatives/friends (loan taken)	2,51,053	3	1,97,375	5
Relatives/friends (outstanding)	2,32,143	3	2,42,865	6
Traders (loan taken)	74,444	2	1,65,000	2
Traders (outstanding)	1,21,250	1	3,54,000	1
Others (loan taken)	3,51,429	1	2,24,053	2
Others (outstanding)	2,31,875	3	2,96,739	3
Total loan taken (formal)	1,82,784	42	2,09,050	39
Total outstanding (formal)	1,99,031	48	2,15,038	48
Total loan taken (informal)	3,01,761	29	2,90,993	31
Total outstanding (informal)	2,84,556	31	3,23,679	34

Note: % indicates the proportion of households within each category.

COMPARISON AMONG STATES

In all three states, most farmers got loans from commercial banks, with money lenders as a backup. In 2020–2021, Andhra Pradesh farmers got the most loans from banks because they needed constant financial support to sustain their livelihoods. On the contrary, Telangana and Karnataka farmers relied more on money lenders because they had less land and grew less profitable crops (Table 8.17). However, things took a turn in 2021–2022. Andhra Pradesh and Karnataka farmers used banks less, while Telangana farmers borrowed more (Table 8.18). This might be because of the government's initiatives to promote and support the farming community in Telangana.

Table 8.19 presents the total loans taken (both formal and informal together) and outstanding amounts in 2020–2021 and 2021–2022. Indebted farmers are stable at 61%; in Karnataka indebted farmers increased from 44% to 59%, while in Andhra Pradesh they increased from 47% to 69%. The average outstanding loan was Rs. 3.29 lakh in Karnataka, followed by Andhra Pradesh (Rs. 3.26 lakh) and Telangana (Rs. 3.12 lakh). The average outstanding loan size was higher among OBC and FC households, large landholding farmers, and irrigated farmers. The severity of debt burden relative to their assets is particularly higher among tenant farmers, as they don't have land, which contributes to more than 70% of assets in rural areas.

Overall, it was really sad to see how loans affected farmers' lives and made us realize the gravity of the situation. Groups with higher debt-to-asset ratios, like poor households, small landowners, tenant farmers, and certain social groups, are at greater risk of financial trouble because they depend more on debt compared to the value of what they own. This makes them more vulnerable to problems like crop failures, changes in market prices, or increasing interest rates. Since they have fewer

TABLE 8.17
Indebtedness by State in Normal Year (2020–2021)

	State Name					
	Telangana		Karnataka		Andhra Pradesh	
Source of Credit	**Mean**	**%**	**Mean**	**%**	**Mean**	**%**
Commercial banks (loan taken)	1,18,953	37	3,43,078	21	1,87,261	53
Commercial banks (outstanding)	1,29,511	29	3,67,054	14	1,91,863	33
Regional rural bank (loan taken)	1,01,576	14	1,92,389	19	1,47,222	13
Regional rural bank (outstanding)	98,300	13	1,98,917	13	1,49,371	7
Co-operative societies/banks (loan taken)	1,18,642	15	1,16,013	17	1,89,154	8
Co-operative societies/banks (outstanding)	1,18,104	10	1,07,688	10	1,91,100	6
SHGs (loan taken)	82,128	8	58,750	1	52,429	7
SHGs (outstanding)	71,286	6	46,250	1	86,667	1
Money lender (loan taken)	2,91,221	19	2,59,595	15	2,08,562	15
Money lender (outstanding)	3,00,843	15	2,81,639	13	2,34,268	9
Relatives/friends (loan taken)	2,41,795	8	2,24,595	8	1,95,833	3
Relatives/friends (outstanding)	2,64,643	6	2,27,500	3	2,50,714	1
Traders (loan taken)	3,04,000	4	3,90,000	1	72,500	1
Traders (outstanding)	4,86,000	1	40,000	0.2	2,00,000	0.2
Others (loan taken)	3,66,000	1	1,75,000	0.2	2,60,636	5
Others (outstanding)		0	1,75,000	0.2	2,05,444	2
Total loan taken (formal)	1,27,190	65	2,44,368	53	1,99,588	69
Total outstanding (formal)	1,27,751	52	2,42,044	35	2,06,498	42
Total loan taken (informal)	3,00,603	31	2,54,872	24	2,31,673	21
Total outstanding (informal)	2,99,991	22	2,68,481	16	2,35,246	12

Note: % indicates the proportion of households within each category.

TABLE 8.18
Indebtedness by State in Bad Year, 2021–2022

	State Name					
	Telangana		Karnataka		Andhra Pradesh	
Source of Credit	**Mean**	**%**	**Mean**	**%**	**Mean**	**%**
Commercial banks (loan taken)	1,92,417	22	2,68,404	10	1,87,047	36
Commercial banks (outstanding)	1,78,690	28	2,89,844	16	1,91,929	47
Regional rural bank (loan taken)	1,35,612	15	2,19,000	6	1,61,930	12
Regional rural bank (outstanding)	1,30,132	15	2,10,100	10	1,60,391	13
Co-operative societies/banks (loan taken)	1,33,056	8	1,16,885	11	2,73,621	6
Co-operative societies/banks (outstanding)	1,74,643	9	1,15,943	11	2,62,559	7
SHGs (loan taken)	98,846	3	52,500	1	69,780	9
SHGs outstanding)	87,750	3	46,250	1	39,412	4

(Continued)

TABLE 8.18 (*Continued*)
Indebtedness by State in Bad Year, 2021–2022

	State Name					
	Telangana		Karnataka		Andhra Pradesh	
Source of credit	**Mean**	**%**	**Mean**	**%**	**Mean**	**%**
Money lender (loan taken)	2,92,009	23	3,72,586	24	2,72,661	23
Money lender (outstanding)	3,13,409	25	3,52,132	27	2,83,475	25
Relatives/friends (loan taken)	2,47,174	5	2,22,857	4	1,53,333	3
Relatives/friends (outstanding)	2,89,962	6	2,06,957	5	2,10,588	4
Traders (loan taken)	1,95,000	2	92,000	1	71,250	2
Traders (outstanding)	3,66,000	1	4,12,000	1	77,500	2
Others (loan taken)	30,000	0	2,37,000	1	2,75,100	4
Others (outstanding)	4,42,000	1	2,17,000	1	2,49,655	6
Total loan taken (formal)	1,74,900	43	1,96,754	27	2,19,456	49
Total outstanding (formal)	1,74,407	50	2,33,054	35	2,22,887	59
Total loan taken (informal)	2,82,029	30	3,39,349	30	2,62,703	30
Total outstanding (informal)	3,23,952	32	3,31,242	34	2,75,399	34

Note: % indicates the proportion of households within each category.

TABLE 8.19
Total Loans Taken and Outstanding Amount in Normal and Bad Years (Mean in Rs. 1,000) with Debt-to-Asset Ratio

			2020–2021		2021–2022	
			Loan Taken	**Outstanding**	**Loan Taken**	**Outstanding**
State	Telangana	Mean	238	219	288	312
		%	74	61	56	61
		DAR	4.1	3.8	5	5.4
	Karnataka	Mean	294	294	316	329
		%	65	44	50	59
		DAR	3.7	3.7	4	4.1
	AP	Mean	249	247	309	326
		%	75	47	61	69
		DAR	5	4.9	6.2	6.5
Religion	Hindu	Mean	261	252	309	326
		%	72	51	56	64
		DAR	4.1	4	4.9	5.1
	Muslim	Mean	198	211	201	209
		%	54	46	46	56
		DAR	4.3	4.6	4.4	4.5
	Christian	Mean	218	206	181	312
		%	65	48	43	43
		DAR	7.5	7.1	6.3	10.8

(*Continued*)

TABLE 8.19 (*Continued*)
Total Loans Taken and Outstanding Amount in Normal and Bad Years (Mean in Rs. 1,000) with Debt-to-Asset Ratio

			2020–2021		2021–2022	
			Loan Taken	Outstanding	Loan Taken	Outstanding
Social group	SC	Mean	181	179	210	234
		%	69	43	54	60
		DAR	6	5.9	7	7.7
	ST	Mean	188	207	210	206
		%	67	51	62	67
		DAR	4.7	5.2	5.3	5.2
	OBC	Mean	274	265	343	362
		%	69	52	57	64
		DAR	3.9	3.8	4.9	5.2
	OC	Mean	278	260	313	334
		%	74	50	53	61
		DAR	4.1	3.8	4.6	4.9
Owned land	Marginal	Mean	172	169	239	245
		%	61	45	53	63
		DAR	9.5	9.3	13.2	13.5
	Small	Mean	183	174	234	235
		%	69	52	53	61
		DAR	6.6	6.3	8.4	8.5
	Medium	Mean	209	204	269	280
		%	73	51	56	62
		DAR	4.6	4.5	6	6.2
	Semi-medium	Mean	308	295	341	359
		%	73	51	57	64
		DAR	3.9	3.8	4.4	4.6
	Large	Mean	593	596	582	691
		%	77	53	59	68
		DAR	2.3	2.4	2.3	2.7
Cultivated land	Marginal	Mean	143	127	192	174
		%	63	50	44	54
		DAR	7.9	7	10.6	9.6
	Small	Mean	143	142	222	213
		%	66	51	49	58
		DAR	5.2	5.1	8	7.7
	Medium	Mean	176	182	204	220
		%	74	52	52	58
		DAR	3.9	4	4.5	4.9
	Semi-medium	Mean	274	263	300	307
		%	71	49	60	66
		DAR	3.5	3.4	3.8	3.9

(*Continued*)

TABLE 8.19 (*Continued*)
Total Loans Taken and Outstanding Amount in Normal and Bad Years (Mean in Rs. 1,000) with Debt-to-Asset Ratio

			2020–2021		2021–2022	
			Loan Taken	Outstanding	Loan Taken	Outstanding
	Large	Mean	495	473	547	608
		%	73	52	61	72
		DAR	2	1.9	2.2	2.4
Poverty status	Non-poor	Mean	252	245	310	319
		%	72	50	59	70
		DAR	3.6	3.5	4.4	4.5
	Poor	Mean	267	256	297	328
		%	70	52	51	55
		DAR	5.7	5.5	5.7	6.2
Tenancy status	Owner	Mean	262	253	305	322
		%	72	51	55	63
		DAR	4	3.9	4.7	5
	Tenant	Mean	187	168	289	320
		%	60	35	64	68
		DAR	14.6	13.1	22.6	25
Irrigation status	Rainfed	Mean	234	227	293	297
		%	72	48	56	62
		DAR	5.2	5	6.5	6.5
	Irrigated	Mean	275	263	311	338
		%	70	52	55	64
		DAR	3.8	3.6	4.3	4.6
	Total	Mean	259	250	304	322
		%	71	51	55	63
		DAR	4.2	4	4.9	5.2

Note: % indicates the proportion of households within each category.

assets and more debt, even small financial issues can cause big problems. They might struggle to pay back loans, and, in the worst cases, they could lose their assets, such as land or property.

We met Masud, a marginal farmer in Karnataka who left agriculture and moved to other non-farm work after his father's death left the family in debt. As a lower-caste member of the village, he also needed to pay a huge dowry for his daughter's marriage. When he discovered that he would inherit his mother's fixed deposit after her death, he wished her death. This was his grievous dilemma: his mother's life or his daughter's life. Finally, he couldn't take his mother's life, but he lost his daughter, who committed suicide after being humiliated by the society of being unmarried at 25.

This story reminds us of the Indian short film *Amma Meri (My Mother)*, which attempted to bridge the gap between imagined and real indebtedness in India, highlighting the dark side of rural poverty. Poverty and indebtedness might make us feel sympathy or even empathy, but we still don't fully understand what poverty actually means and how it affects those who live in that situation.

The farmers borrowed money to make life better but ended up in debt, struggling to make ends meet. Their debt was not just a number; it was a reality that affected their daily lives and future.

REFERENCES

Arora S. (2021). *A Study of Land Distribution, Tenancy and the Land Market in a Central Indian Village, Monograph21/1*. Society for Social and Economic Research, New Delhi.

Asian Development Bank (ADB). (2020). COVID-19 and Rural Poverty in Southeast Asia: An Impact Assessment. *ADB Briefs*, 151.

Babu KK. (2018). Agrarian Distress in Andhra Pradesh (A Study on Guntur District). *International Journal of Business, Management and Allied Sciences (IJBMAS)*, 5(S2), pp. 65–71.

Binswanger-Mkhize HP. (2012). Is There Too Much Hype about Index-Based Agricultural Insurance? *Journal of Development Studies*, 48(2), pp. 187–200.

Chand R, Srivastava SK, Singh J. (2017). *Changing Structure of Rural Economy of India: Implications for Employment and Growth, NITI AAYOG Discussion Paper*. Niti Aayog.

Chavas JP, Di Falco S, Adinolfi F, Capitanio F. (2019). Weather Effects and Their Long-Term Impact on the Distribution of Agricultural Yields: Evidence from Italy. *European Review of Agricultural Economics*, 46(1), pp. 29–51.

Chhikara KS, Kodan AS. (2014). Farmers' Indebtedness in Haryana: A Study. *Journal of Rural Development*, ,32(4), pp. 347–365.

Darling M. (1925). *The Punjab Peasant in Prosperity and Debt*. Oxford University Press.

Dev SM, Sengupta R. (2020). *Impact of COVID-19 on the Indian Economy: A Preliminary Assessment*. Indira Gandhi Institute of Development Research (IGIDR).

Food and Agriculture Organization (FAO). (2021). *The Impact of COVID-19 on Food and Agriculture in Latin America and the Caribbean*. FAO, Rome.

Hoang VN, Nghiem SH, Wilson C. (2023). Social Networks with Organisational Resource, Generalised Trust and Informal Loans: Evidence from Rural Vietnam. *Economic Analysis and Policy*, 77, pp. 388–402.

Isakson R. (Ed.). (2016). *New Directions in Agrarian Political Economy: Global Agrarian Transformations*, Volume 1 (1st ed.). Routledge. https://doi.org/10.4324/9781315689661

Jodhka SS. (1995). Bureaucratisation, Corruption and Depoliticisation: Changing Profile of Credit Co-Operatives in Rural Haryana. *Economic & Political Weekly*, 30(1), pp. 53–56.

Karthick V, Madheswaran S. (2018). Access to Formal Credit in the Indian Agriculture: Does Caste Matter? *Journal of Social Inclusion Studies*, 4(2), pp. 169–195.

Kumar KS. (2018). Farmer, Wife, Two Kids Commit Suicide in Karnataka over Debt. *The New Indian Express*. https://www.newindianexpress.com/states/karnataka/2018/Sep/23/farmer-wife-two-kids-commit-suicide-in-karnataka-over-debt-1875983.html

Kumar SM. (2013). Does Access to Formal Agricultural Credit Depend on Caste? *World Development*, 43(3), pp. 315–328.

Lavanya T. (2016). Agricultural Indebtedness in Andhra Pradesh. *Indian Journal of Economics and Development*, 12(1a), pp. 323–328.

Mellor JW. (2017). *Agricultural Development and Economic Transformation: Promoting Growth with Poverty Reduction.* Springer (Palgrave Studies).

Narayanan S. (2016). The Productivity of Agricultural Credit in India. Agricultural *Economics*, 47(4), pp. 399–409.

National Sample Survey Office. (2014). *Key Indicators of Debt and Investment in India, NSS 70th Round, (January*–December 2013*).* Report No. NSS KI (70/18.2). Ministry of Statistics and Programme Implementation, National Sample Survey Office, New Delhi.

NSS. (2013). https://microdata.gov.in/nada43/index.php/catalog/133

Pandey GK. (2016). Extent, Magnitude and Determinants of Indebtedness among Farmers in Eastern India: A Survey-Based Study. *Indian Journal of Agricultural Economics*, 71(4), pp. 450–462.

Ramprasad V. (2019). Debt and Vulnerability: Indebtedness, Institutions and Smallholder Agriculture in South India. *The Journal of Peasant Studies*, 46(6), pp. 1286–1307.

Rao S. (2017). Caste Discrimination and Agricultural Performance in India. *Economic and Political Weekly*, 52(25–26), pp. 32–38.

Rao Narasimha P, Suri KC. (2006). Dimensions of Agrarian Distress in Andhra Pradesh. *Economic and Political Weekly*, 41(16), pp. 1546–1552.

Reddy AA, Raju SS, Bose A. (2020). Indebtedness and Agrarian Distress in India. *The Microfinance Review*, 12(1), pp. 20–38.

Reddy DN, Mishra S. (2012). Agrarian Crisis in India. Oxford Academic (Delhi, 2010; online edn, 18 Oct. 2012). https://doi.org/10.1093/acprof:oso/9780198069096.001.0001.

Sajjad H, Chauhan C. (2012). Agrarian Distress and Indebtedness in Rural India: Emerging Perspectives and Challenges Ahead. *Journal of Geography and Regional Planning*, 5(15), p. 397.

Singh J. (2006). Changing Agrarian Relationships in Rural Areas. *Indian Journal of Agricultural Economics*, 61(1), pp. 81–84.

Singh S, Bhogal S, Singh R. (2014). Magnitude and Determinants of Indebtedness among Farmers in Punjab. *Indian Journal of Agricultural Economics*, 69(2), pp. 243–256.

Sreenivasulu Y. (2023). Forced Tenancy as a Trajectory of Tenant Farmers' Suicides: A Study of Two Mandals in Nalgonda District (Telangana). *Indian Journal of Agricultural Economics*, 78(2), pp. 186–201.

Torhonen MP, Prettitore PS, Scheierling SM, Hilhorst T, Roquet V, Zakout W. (2019). *Enabling the Business of Agriculture 2019.* World Bank Group.

Tran MC, Gan CEC, Hu B. (2016). Credit Constraints and Their Impact on Farm Household Welfare: Evidence from Vietnam's North Central Coast Region. *International Journal of Social Economics*, 43(8), pp. 782–803.

9 Investigating Crop Losses and Recovery

None of our team members wanted to believe that loans for farming were the sole cause of farmers' debts. We felt there was something more. After multiple rounds of discussions with farmers and local government members, and reflecting on our own instincts, we ensured that repeated crop failures were another major reason for their indebtedness. Crop failures stemming from various factors have compelled farmers to increase borrowing to cover their expenses and needs (Mishra, 2008; Ramprasad, 2019). Among the primary causes of these failures are climatic extremes and biological threats. Changing seasons, characterized by a range of climatic factors such as floods, cyclones, storms, heatwaves, droughts, and severe pest and disease attacks, have rendered farming increasingly risky (Jat et al., 2020; Oerke, 2006). For instance, dry summers lead to drought conditions that can stress crops, causing them to wither and die under intense heat (Im et al., 2017). Conversely, excessive rainfall during the monsoon season can result in flooding and waterlogging, which may wash away crops or create conditions conducive to the proliferation of diseases (Kim et al., 2023). Pests and diseases, exacerbated by climate change, have become more severe and frequent, further compounding the risks for farmers (Frank, 2024; Reddy et al., 2024). Many farmers identify these climatic and biological factors as primary contributors to their crop losses, illustrating the growing unpredictability and vulnerability of agriculture (Haque and Khan, 2017). We got to know this from the farmers when one of them stated:

I lost all my investment this season, but what worries me more is the next season. I don't have any land left to grow crops, and the water is slowly receding. It feels like I won't be able to grow any crops at all.

Another farmer Ganesh Kumar said that after the COVID-19 pandemic, he was hopeful that 2021–2022 would be better, but the extreme weather caused a huge financial loss. "*In March, an unexpected temperature rise hit just as the wheat crop was maturing*," he said, "*and it led to the drastic fall in the production*."

But it was not just the weather that was causing the losses. Pests like locusts, aphids, bollworms, and fungi were also destroying the crops. The farmers did not have access to good-quality pesticides or the knowledge of proper pest control practices. It was a losing battle for them.

Not only that, there was another battle going on under the ground that we discovered during our investigation. The use of chemical fertilizers was weakening the soil and causing erosion (Bhattacharyya et al., 2015). This made it even more difficult for crops to grow properly, leading to further losses. The issues did not end with harvest. Due to the lack of suitable storage facilities, the crops would decay and go to waste. However, there were some glimmers of hope. We met a few farmers who realized the seriousness of the situation and by adopting organic farming methods were able to minimize their losses to a certain extent.

DOI: 10.1201/9781003582878-9

DEFENCES AGAINST CROP LOSSES

When the farmers were trying to find ways to mitigate it, the government stepped in with some initiatives to tackle this issue. The Indian government introduced crop insurance schemes like the Pradhan Mantri Fasal Bima Yojana (PMFBY) to protect farmers from disasters like floods, pests, and diseases. The government also invested more money in research, allowing scientists to work harder and come up with better ways to fight against pests and diseases. The government also focused on improving irrigation systems by constructing canals and enhancing drip irrigation and sprinkler systems to make sure farmers always get sufficient water for their crops. But that was not all. They also developed cold storage and warehouses to keep crops fresh and safe before marketing and improved transportation services to easily reach the market. Finally, they encouraged methods like conservation agriculture and organic farming to make the farm more resilient in the face of adversity.

However, among all the initiatives taken, crop insurance played the most crucial role in mitigating crop losses. It provided a safety net for farmers in case of any unforeseen circumstances. But, for this to work effectively, farmers needed to understand the specific guidelines and procedures related to the insurance scheme in their respective places. Because of that, the coverage was quite low and fewer farmers chose to get crop insurance (Ghosh et al., 2020; Sona and Muniraju, 2018)). Nevertheless, these government programmes were praised by all, and farmers were finally able to take a breath of relief. Their hard work and dedication were no longer in vain, as their crops were now safer and more protected than ever before. But the reality seemed to be a little different. Even with a lot of research, there was not much evidence that shows traditional crop insurance positively helps farmers (Smith and Watts, 2010; Fadhliani et al., 2019).

WHAT WAS THE REALITY OF CROP LOSS AND HOW DO FARMERS EXPERIENCE INSURANCE CLAIMS?

Although it was difficult, we felt it was important to shed light on the harsh reality faced by these farmers. The real situation we heard from the farmers was terrible. We met several farmers who spoke of the unpredictable weather, the lack of government support, and the devastating losses over the past two years. One farmer showed us his barren fields and tearfully described how he lost everything to a cyclone and pest attack. He had invested all his savings into the crop, hoping to provide support for his family, but now he was left with nothing. Unfortunately, many farmers like him, across the region, suffered this hardship not once but twice with over 74% in Andhra Pradesh alone. However, only some farmers had received insurance claims. It was painful to realize that, despite their resilience, many farmers were not getting the support they desperately needed. This shows that even though crop insurance provides many benefits, its adoption rate in India is still low (Biswal and Bahinipati, 2022). And so, to explore the reality, we continued to do our research among different groups.

Religious Groups

We spoke to farmers of different religious backgrounds, and all of them agreed that crop loss was a common issue among them. In fact, over half of them already experienced crop loss twice in consecutive years (Table 9.1). As we questioned more, we discovered that Christian farmers were more prominent among them. But surprisingly, there were still some farmers who experienced no loss at all. However, there was a bright side. When it came to insurance claims, Christian farmers had the highest success rate in receiving insurance claims, with half of them receiving assistance. So, it became true that religiosity can influence farmers' insurance demand (Sihem, 2019).

Social Groups

Similar to the religious group, the frequency of crop loss also varied among the social groups because caste-based networks existed in Indian villages (Mobarak and Rosenzweig, 2013). More than half of the farmers were hit by crop failure and experienced double losses. Except for ST farmers, these numbers remained consistent across all social groups, with general castes having the highest percentage of loss (Table 9.2). But what was even more alarming was the fact that, while most of the farmers were facing crop loss, only a small percentage had actually claimed insurance. Interestingly, general farmers also had the highest level of comfort with insurance coverage, with 40% expressing confidence in it. Chander et al. (2020) also found similar findings in their study. This was because the general caste farmers were more aware of crop insurance than others (Mukherjee and Pal, 2019). This showed a huge disparity among the different social groups and their access to crop insurance.

Landholding Groups (Own Land)

A group of farmers from different landholding groups surrounded us while we were interviewing. We could see the worry and stress on the faces of those who had faced many challenges in the past few years. We started by asking them about their

TABLE 9.1
Crop Loss and Insurance Claims in the Last Two Years by Religion Category of Farmers

Frequency of Crop Loss in the Last Two Years	Hindu	Muslim	Christian	Total (%)
0	9.2	15.4	4.3	9.4
1	36.7	23.1	21.7	36
2	54.1	61.5	73.9	54.6
a. How many times have you made a loss on a production cycle in the last two years?	1.4	1.5	1.7	1.5
b. Did you get claims from the insurance company for your loss?	30%	20%	50%	30%

TABLE 9.2
Crop Loss and Insurance Claims in the Last Two Years by Social Group

Frequency of Crop Loss in the Last Two Years	SC	ST	OBC	General (%)
0	7.6	13.7	9.1	9.0
1	32.8	47.9	39.5	30.3
2	59.5	38.4	51.4	60.8
a. How many times have you made a loss on a production cycle in the last two years?	1.5	1.2	1.4	1.5
b. Did you get claims from the insurance company for your loss?	20%	10%	10%	40%

experiences with crop loss. To our surprise, we found that the highest percentage of losses were among large farmers, with almost 60% of them having suffered crop loss twice in the past two years (Table 9.3). Despite the size of their land, the average number of times a farmer had lost their crop in two years was almost the same. Also, when we talked about crop insurance, we found that medium and large farmers had higher rates of benefitting from claims. This was because the process and requirements for claiming insurance were more complex, making it harder for small and marginal farmers to access it.

Land Holding Groups (Total Cultivated Land)

We observed that losses varied depending on the size of total cultivated land, with large farmers experiencing maximum losses. These losses were not just impacting farmers' income but also their mental and emotional well-being. Many of them had taken loans to invest in their crops and were now struggling to repay them due to these losses. It was a relief to see that claims were highest among medium and large farmers (Table 9.4), offering them some hope to recover some of their losses and keep continuing their livelihood.

TABLE 9.3
Crop Loss in the Last Two Years by Own-Land Size Category

Frequency of Crop Loss in the Last Two Years	Marginal	Small	Semi-Medium	Medium	Large (%)
0	5.5	11	10.1	8.8	11.9
1	42.7	34.1	37.6	33.8	28.7
2	51.8	54.9	52.3	57.4	59.4
a. How many times have you made a loss on a production cycle in the last two years?	1.5	1.4	1.4	1.5	1.5
b. Did you get claims from the insurance company for your loss?	20%	20%	20%	30%	30%

TABLE 9.4
Crop Loss in the Last Two Years by Land-Cultivated Size Category

Frequency of Crop Loss in the Last Two Years	Marginal	Small	Semi-Medium	Medium	Large	Total (%)
0	10	12.8	10.8	7.3	8.2	9.4
1	47.5	37.9	39.8	34.4	27.2	36
2	42.5	49.2	49.4	58.3	64.7	54.6
a. How many times have you made a loss on a production cycle in the last two years?	1.3	1.4	1.4	1.5	1.6	1.5
b. Did you get claims from the insurance company for your loss? (% of farmers who paid insurance premium)	20	10	20	30	30	30

Poverty Groups

However, in the villages where poverty was already high, it was strange to see even the non-poor farmers taking the biggest hit with their losses. There was a big gap between the non-poor and poor farmers, who were dealing with losses for two consecutive years. But what really concerning was that the insurance claims did not change between these two years (Table 9.5). It was like the insurance companies were not giving enough support to these struggling farmers.

Tenancy Groups

The situation was further highlighted when we compared tenant farmers to owner farmers. Despite battling with higher crop losses, the owner farmers were getting more benefits even though they were dealing with more crop losses (Table 9.6). We heard about this insurance company covering crop loss, but only 20% of the tenant farmers received any compensation from it. That made us think: why were not more farmers getting these insurance benefits? So, we decided to further speak to the farmers and gather firsthand information. To our surprise, we found out that only 20%

TABLE 9.5
Crop Loss in the Last Two Years by Poverty Status

Frequency of Crop Loss in the Last Two Years	Poverty status, 2021			Poverty status, 2022		
	Poor	Non-poor	Total	Poor	Non-poor	Total (%)
0	10.4	7.2	9.4	12.2	6.2	9.4
1	41	25.8	36	36.2	35.8	36
2	48.5	66.9	54.6	51.7	58	54.6
a. How many times have you made a loss on a production cycle in the last two years?	1.4	1.6	1.5	1.4	1.5	1.5
b. Did you get claims from the insurance company for your loss? (% of farmers who paid insurance premium)	20	30	30	20	30	30

TABLE 9.6
Crop Loss in the Last Two Years by Tenancy Status

Frequency of Crop Loss in the Last Two Years	Owner Farmer	Tenant Farmer	Total (%)
0	9.6	5.6	9.4
1	35.8	40.3	36
2	54.7	54.2	54.6
a. How many times have you made a loss on a production cycle in the last two years?	1.5	1.5	1.5
b. Did you get benefits from the insurance company for your loss? (% of farmers who paid insurance premium)	30	20	30

of them had actually applied for the claim, which means the participation of tenant farmers was very low (Jayathilaka and Abeynayaka, 2013). But what was even more strange was the fact that the few farmers who did apply for the claim had no problem getting the claims. It was also true that certain requirements like land records exclude tenant farmers from participating in crop insurance like PMFBY (Ghosh et al., 2020). But it was not like the insurance company was neglecting; we figured out that it was mainly the lack of awareness to take advantage of this awesome policy.

Irrigation Groups

But the disparities did not end there. We dug it up that the farmers with dryland had more crop loss (once and twice) with a total of 93% of farmers compared to irrigated farmers with 89% (Table 9.7). However, the benefits obtained from insurance claims were highest for dryland farmers, which was against the common perception that irrigated land had better crop yields and more profits.

COMPARISON AMONG STATES

In both years, Karnataka had the highest losses among the states. When we visited farms and spoke to farmers, we found out it was mainly because of problems like poor irrigation, unpredictable weather, and pest infestation. It was disheartening to

TABLE 9.7
Crop Loss in the Last Two Years for Dryland and Irrigated Farmers

Frequency of Crop Loss in the Last Two Years	Dry Land	Irrigated Land	Total (%)
0	6.6	11.1	9.4
1	25.6	42.5	36
2	67.8	46.5	54.6
a. How many times have you made a loss on a production cycle in the last two years?	1.6	1.4	1.5
b. Did you get benefits from the Insurance Company for your loss? (% of farmers who paid insurance premium)	40	10	30

see the hard work of these farmers go to waste because of things they could not control. Andhra Pradesh had the lowest losses compared to the other states (Table 9.8). Upon further investigation, we figured out it was because they used better farming methods and had a stronger support system for their farmers.

STORIES OF RESILIENCE IN CROP INSURANCE

The big problem for farmers in India is that they do not have much money to handle the risk of crop failures. That was why crop insurance was crucial. The year 2021–2022 showed a significant increase in crop loss, with chillies being the most affected crop in all three states (Table 9.9). However, what caught our attention was the difference in the number of claims made by farmers in different states. In our research, we came across two different crop insurance schemes: the YSR zero premium crop insurance scheme and the PMFBY. While both schemes aim to provide insurance coverage to farmers, their approaches and outcomes are vastly different.

TABLE 9.8
Crop Loss in the Last Two Years by States

Frequency of Crop Loss in the Last Two Years	Telangana	Karnataka	Andhra Pradesh	Total (%)
0	11.5	15.4	1.3	9.4
1	44.1	39.8	24.4	36
2	44.3	44.8	74.4	54.6
a. How many times have you made a loss on a production cycle in the last two years?	1.3	1.3	1.7	1.5
b. Did you get benefits from the Insurance Company for your loss? (% of farmers who paid insurance premium)	0	0	70	30

TABLE 9.9
Insurance Claims under Different Crops

State Name	Crops		Loss Claimed in 2020–2021 (Rs.)	Claim Amount Received in 2020–2021 (Rs.)	Loss Claimed in 2021–2022 (Rs.)	Claim Amount Received in 2021–2022 (Rs.)
Telangana	Cotton	Mean	20,000		75,000	
		N	1		1	
	Chillies	Mean	30,000			
		N	1			
	Total	Mean	25,000		75,000	
		N	2		1	

(Continued)

TABLE 9.9 (*Continued*)
Insurance Claims under Different Crops

State Name	Crops		Loss Claimed in 2020–2021 (Rs.)	Claim Amount Received in 2020–2021 (Rs.)	Loss Claimed in 2021–2022 (Rs.)	Claim Amount Received in 2021–2022 (Rs.)
Karnataka	Cotton	Mean	83,560	30,750	1,01,400	12,000
		N	15	4	5	1
	Paddy	Mean	82,625	13,750	29,500	24,000
		N	8	4	4	1
	Chillies	Mean	30,000		40,000	
		N	3		2	
	Red gram	Mean	33,000	14,320	27,250	
		N	14	5	8	
	Chickpea	Mean	22,500		70,000	
		N	2		1	
	Sorghum	Mean	18,125	12,500	15,000	
		N	8	4	1	
	Others	Mean	17,500		5,15,000	3,20,000
		N	2		2	1
	Total	Mean	51,758	17,624	88,609	1,18,667
		N	52	17	23	3
Andhra Pradesh	Cotton	Mean	1,75,991	91,651	69,070	19,114
		N	56	56	115	117
	Paddy	Mean	26,040	15,900	58,300	32,500
		N	10	10	2	2
	Chillies	Mean	58,000	12,100	1,45,055	96,529
		N	3	3	73	75
	Red gram	Mean	22,219	16,888	80,785	42,872
		N	16	16	79	82
	Maize	Mean	5,000	22,000	0	50,000
		N	2	2	1	1
	Chickpea	Mean			1,50,000	30,000
		N			1	1
	Black gram	Mean	95,000	38,000	35,947	26,671
		N	2	2	19	21
	Others	Mean	39,167	41,667	70,394	59,992
		N	3	3	33	35
	Total	Mean	1,19,162	63,510	87,265	47,295
		N	92	92	323	334
Total	Cotton	Mean	1,54,568	87,591	70,455	19,054
		N	72	60	121	118
	Paddy	Mean	51,189	15,286	39,100	29,667
		N	18	14	6	3

(*Continued*)

TABLE 9.9 (*Continued*)
Insurance Claims under Different Crops

State Name	Crops		Loss Claimed in 2020–2021 (Rs.)	Claim Amount Received in 2020–2021 (Rs.)	Loss Claimed in 2021–2022 (Rs.)	Claim Amount Received in 2021–2022 (Rs.)
	Chillies	Mean	42,000	12,100	1,42,253	96,529
		N	7	3	75	75
	Red gram	Mean	27,250	16,276	75,862	42,872
		N	30	21	87	82
	Maize	Mean	5,000	22,000	0	50,000
		N	2	2	1	1
	Chickpea	Mean	22,500		1,10,000	30,000
		N	2		2	1
	Sorghum	Mean	18,125	12,500	15,000	
		N	8	4	1	
	Black gram	Mean	95,000	38,000	35,947	26,671
		N	2	2	19	21
	Others	Mean	30,500	41,667	95,800	67,215
		N	5	3	35	36
	Total	Mean	93,865	56,354	87,319	47,930
		N	146	109	347	337

N, number of farmers.

Andhra Pradesh had YSR zero premium crop insurance programme, where farmers were provided with insurance coverage without having to pay any premiums. This has been a lifeline for many farmers, especially in bad years when crop losses are high. It had simpler enrolment and claims processes, which led to higher satisfaction among farmers. We heard stories of farmers who were bankrupted due to crop losses, but, thanks to the YSR scheme, they were able to receive timely compensation and continue farming. On the contrary, PMFBY was implemented in Karnataka, which required farmers to pay subsidized premiums. Farmers enrolled in PMFBY expressed their frustration with the complicated procedures for enrolment and claims. They mentioned that, in bad years, they were left with little to no compensation, which puts them in a difficult financial situation. Telangana, unfortunately, had no such crop insurance scheme in place, resulting in a lack of claims made by farmers. Even though cotton growers in the state made claims of 75,000 in the year 2021–2022, they received no compensation as there was no accompanying scheme available.

It was a humbling experience to witness the strengths and determination of these farmers and hard to believe the reality of the lives of these farmers. As we listened to their stories, we understood that crop loss affected their livelihood and pushed them into debt and poverty. We felt a sense of injustice towards the farmers who were facing such disparities and inequalities. Also, after discovering the stories of resilience in crop insurance, it was clear that we need to find ways of improving agricultural insurance schemes locally and nationally (Reddy et al., 2021). So the whole scenario highlighted the need for specialized support systems to uplift the agricultural community, especially those with limited resources.

REFERENCES

Bhattacharyya R, Ghosh BN, Mishra PK, Mandal B, Rao CS, Sarkar D, Das K, Anil KS, Lalitha M, Hati KM, Franzluebbers AJ. (2015). Soil Degradation in India: Challenges and Potential Solutions. *Sustainability*, 7, pp. 3528–3570.

Biswal D, Bahinipati CS. (2022). Why Are Farmers Not Insuring Crops against Risks in India? A Review. *Progress in Disaster Science*, 15, 100241.

Chander S, Kumari V, Tyagi R. (2020). Socio-Economic Factors Affecting the Crop Insurance Scheme in Haryana – A Sociological Analysis. *Journal of Pharmacognosy and Phytochemistry*, 9(2), pp. 42–45.

Fadhliani Z, Luckstead J, Wailes EJ. (2019). The Impacts of Multiperil Crop Insurance on Indonesian Rice Farmers and Production. *Agricultural Economics*, 50, pp. 15–26.

Frank EG. (2024). The Economic Impacts of Ecosystem Disruptions: Costs from Substituting Biological Pest Control. *Science*, 385(6713), eadg0344.

Ghosh RK, Gupta S, Singh V, Ward PS. (2020). Demand for Crop Insurance in Developing Countries: New Evidence from India. *Journal of Agricultural Economics*, 72(1), pp. 293–320.

Haque MI, Khan MR. (2017). Farmers' Sensitivity to Crop Loss: Evidence from India. *International Journal of Economic Research*, 14(8), pp. 91–100.

Im ES, Pal JS, Eltahir EA. (2017). Deadly Heat Waves Projected in the Densely Populated Agricultural Regions of South Asia. *Science Advances*, 3(8), e1603322.

Jat ML, Chakraborty D, Ladha JK, Rana DS, Gathala MK, McDonald A, Gerard B. (2020). Conservation Agriculture for Sustainable Intensification in South Asia. *Nature Sustainability*, 3(4), pp. 336–343.

Jayathilaka DPDM, Abeynayaka NR. (2013). Assessing Farmers' Perception, Awareness and Influential Factors to Purchase a Crop Insurance as a Tool for Risk Management. In *Proceedings of 12th Agricultural Research Symposium*, Makandura, Srilanka, 2013, pp. 394–398.

Kim W, Iizumi T, Hosokawa N, Tanoue M, Hirabayashi Y. (2023). Flood Impacts on Global Crop Production: Advances and Limitations. *Environmental Research Letters*, 18(5), 054007.

Mishra S. (2008). Risks, Farmers' Suicides and Agrarian Crisis in India: Is There a Way Out?. *Indian Journal of Agricultural Economics*, 63(1).

Mobarak AM, Rosenzweig MR. (2013). Informal Risk Sharing, Index Insurance, and Risk Taking in Developing Countries. *American Economic Review: Papers and Proceedings*, 103(3), pp. 375–380.

Mukherjee S, Pal P. (2019). On Improving Awareness about Crop Insurance in India. *Review of Agrarian Studies*, 9(1), pp. 46–68.

Oerke EC. (2006). Crop Losses to Pests. *Journal of Agricultural Science*, 144(1), pp. 31–43.

Ramprasad V. (2019). Debt and Vulnerability: Indebtedness, Institutions and Smallholder Agriculture in South India. *The Journal of Peasant Studies*, 46(6), pp. 1286–1307.

Reddy AA, Bhattacharya A, Reddy SV, Ricart S. (2021). Farmers' Distress Index: An Approach for an Action Plan to Reduce Vulnerability in the Drylands of India. *Land*, 10(11), p. 1236.

Reddy AA, Reddy M, Mathur V. (2024). Pesticide Use, Regulation, and Policies in Indian Agr iculture. *Sustainability*, 16(17), 7839.

Siuhem E. (2019). Economic and Socio-Cultural Determinants of Agricultural Insurance Demand across Countries, *Journal of the Saudi Society of Agricultural Sciences*, 18(2), pp. 177–187.

Smith VH, Watts M. (2010). The New Standing Disaster Program: A SURE Invitation to Moral Hazard Behavior. *Applied Economic Perspectives and Policy*, 32, pp. 154–169.

Sona HC, Muniraju Y. (2018). Crop Insurance: Farmers Perception and Awareness – A Study with Special Reference to Kodagu District. *International Journal of Research in Economics and Social Sciences*, 8(2), pp. 222–231.

10 Barriers to Adopting Antifragile Farming Practices

Gradually we learned and accepted how bad weather makes farming harder. It might make staple crops produce up to 30% less because of the reduced productivity and sometimes the failure of the crop (Jain et al., 2015). Despite such challenges, farmers never give up. Instead, they showed courage, came together, and tried to find solutions in difficult times. Because they believe, adapting to climate change is the key approach to prevent its severe impacts (Scheffran et al., 2012). As the researchers, we also spent years studying different farming methods and techniques, looking for ways to make it more sustainable and resilient in the face of changing weather patterns. But these farmer's constant determination and courage motivated us to continue our research and find ways to support them. Many farmers stated that they strongly felt that making their farms resilient to climate change could improve their lives, ensuring not only their daily food but also a better future (Acevedo et al., 2020). So they discovered some simple but reliable techniques and also tried to adopt the methods initiated by the government that could bring their hope back.

We met Ramulu, a 45-year-old farmer in Telangana, who had to give up farming due to a lack of water for irrigation. "*I was a farmer with no crops*," said Ramulu. "*The rains used to nurture my land, but now they are scanty and unpredictable*," he explained. "*When the weather changed, I installed a bore well, but it dried up. The whole purpose was lost*," he recalls. Despite the challenges, Ramulu kept investing in his land, hoping for a good monsoon. However, his expenses grew with his family's emergency. "*All I got in return were losses*," he laments. Eventually, Singh turned to daily wage work, which is inconsistent. "*But after knowing about some special practices from some NGO workers I was given back the hope*," he sighs.

Ramulu's story pointed out one of the biggest challenges that Ramulu and other farmers faced was the lack of water during droughts. To combat this, they started learning about "water-saving crops" like millets, pulses, and oilseeds that require less water (Chazovachii et al., 2012). These crops became a good support during drought. They also adopted conservation agriculture, a "soil-friendly farming method," to deal with climate change. Instead of deep ploughing, they tilled the soil just a little, covered it with mulch, and planted different crops together

 DOI: 10.1201/9781003582878-10

to keep the soil healthy and moist. It did not end here. To add to their income, farmers started planting trees on their farms through "agroforestry." Trees not only protect against harsh weather conditions but also bring in additional income (Taillandier et al., 2023; Mulugeta, 2014). To save water, farmers switched to "improved irrigation systems" like drip irrigation and sprinklers. These methods were super-efficient at watering their crops, even in challenging conditions. Even their animals got a transformation with "climate-resilient livestock management." It provided better nutrition, improved breeding, and disease control, making the animals stronger against natural disasters (Pankaj et al., 2024). Together, these techniques and methods could change farmers' fortune and help them overcome natural challenges (Figure 10.1). But what if there was a calamity and their crops were destroyed? That was where "weather-based crop insurance" came to the rescue. By using the weather information, it helped farmers when things went wrong (Vishnoi et al., 2020).

However, adopting these new techniques was not that easy, particularly for small farmers. From 2020 to 2022, farmers faced numerous challenges in trying new ways to protect their crops. They found that the supplies were very expensive, and borrowing money made them trapped. They also struggled with poor infrastructure and unreliable electricity and internet, making it difficult to access information about these techniques. There were other troubles too by some private companies, which forced the farmers to buy seeds they did not need. This added financial pressure on already struggling farmers. Additionally, many farmers were hesitant to take risks with new techniques, fearing failure and societal pressures.

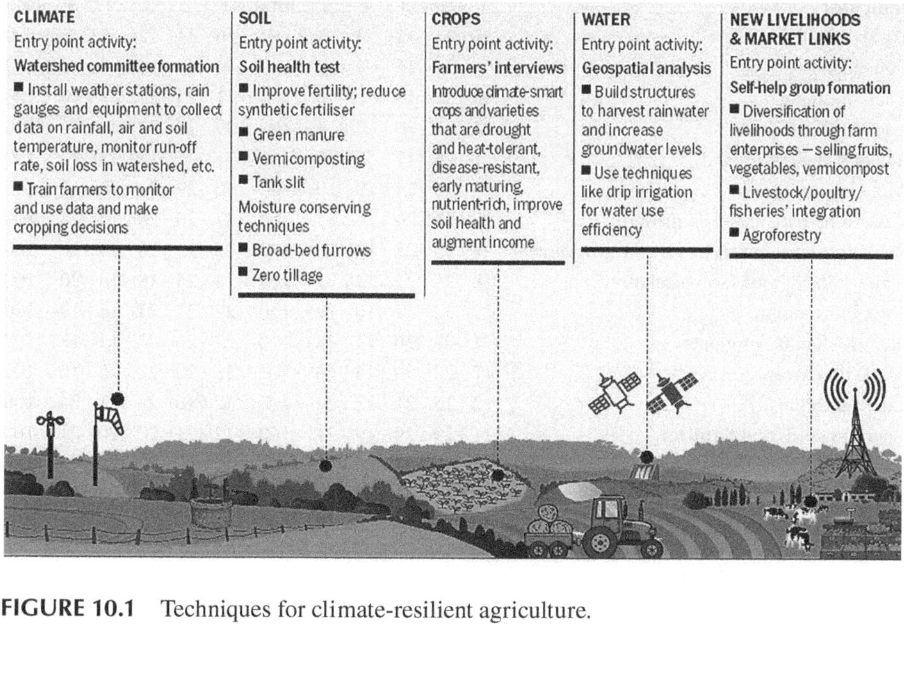

FIGURE 10.1 Techniques for climate-resilient agriculture.

WHAT CHALLENGES DO FARMERS FACE IN ADOPTING NEW MEASURES?

Even though the government put its efforts into helping farmers deal with climate change and protect their crops, the situation was not as hopeful as expected. One farmer, burdened by the high cost of inputs like fertilizers and pesticides rented, said: "*I simply can't afford the necessary inputs anymore. They've become too costly, making it difficult to keep our crops healthy.*"

When we looked around, we saw many farmers facing the same challenges. The lack of money and the fear of being in debt made it hard for them to try any new methods. In addition to this, farmers also dealt with a scarcity of skilled labour and limited access to machinery and extension services. These practical limitations complicated their struggles further (Table 10.1).

But it was not just about the financial aspect. Other factors had a neutral effect on agricultural activities, like the force from private seed companies to purchase their products, low education levels, and the use of spurious seeds and fertilizers. These may not directly affect their farming, but they did add to their overall struggles.

We saw that the farmers wanted to adopt climate-resilient farming, but several crises were holding them back. We feel very sad to recall the framer who expressed his heartfelt regret by saying:

TABLE 10.1
Challenges in the Adoption Measures 2020–2021 and 2021–2022 (% of Farmers Rated Indicators)

	2020–2021						2021–2022					
Indicator	**1**	**2**	**3**	**4**	**5**	**Total**	**1**	**2**	**3**	**4**	**5**	**Total**
Costly inputs	56	24	12	4	4	100	56	24	12	5	3	100
Indebtedness	44	25	16	8	8	100	46	23	14	8	8	100
Lack of capital	45	24	18	8	6	100	41	25	21	8	6	100
Lack of liquid cash	43	27	20	7	4	100	39	27	22	8	4	100
Tenancy (leased-in land)	27	28	17	8	20	100	31	25	17	7	20	100
More labour intensive	25	32	24	10	10	100	30	29	21	10	11	100
Lack of helping hands on the farm	26	21	26	14	14	100	27	20	25	14	14	100
Lack of extension services by the government	24	13	27	18	18	100	25	13	28	16	18	100
Non-availability of farm machinery	18	28	21	13	21	100	24	24	18	13	20	100
Lack of irrigation	22	22	21	16	19	100	24	21	21	16	19	100
Non-availability of inputs	21	28	20	13	18	100	22	26	22	13	18	100
Market far away	17	23	29	14	16	100	18	24	28	15	16	100
Low education	12	15	27	22	24	100	12	16	28	21	24	100
Spurious seeds and fertilizer	11	19	29	19	21	100	12	21	27	20	21	100
Lack of access to information	11	20	31	23	15	100	12	20	32	21	15	100
Low productive soils	7	17	33	23	20	100	8	16	32	24	20	100
Force from private seed companies	4	11	25	23	38	100	4	12	25	21	38	100

1, severe limitation; 5, not severe.

"Farming is not just a job for us, it is our way of life. But if things go like this, we'll have no choice but to give up our land and find other ways to survive."

So we felt that the government needs to understand the plight of farmers and provide them the necessary support and resources to overcome these constraints.

The adoption of antifragile practices—strategies that enable farmers to not only withstand shocks but thrive in the face of them—faces numerous constraints. These constraints are interrelated and often reinforce each other, creating a challenging environment, particularly for smallholders and tenant farmers.

ECONOMIC CONSTRAINTS

One of the most significant barriers to adopting antifragile practices is the economic burden on farmers. High-cost inputs, such as quality seeds, fertilizers, and pesticides, make it difficult for farmers to transition to more resilient agricultural methods (Pingali, 2012). Indebtedness further exacerbates this challenge, as many farmers are already struggling to repay loans (Binswanger-Mkhize, 2012; Reddy, 2012). The situation is compounded by the lack of capital and liquid cash, which limits farmers' ability to invest in more robust farming systems (Barrett et al., 2001). Tenant farmers who lease land face additional challenges, as the uncertainty of land tenure diminishes their incentive to invest in long-term antifragile practices (Bardhan, 1984).

LABOUR AND RESOURCE CONSTRAINTS

Antifragile practices are often more labour-intensive, requiring efforts such as crop diversification, soil improvement, and sustainable pest management (Ellis, 2000). However, many farmers face a shortage of labour due to rural–urban migration and an ageing farming population (FAO, 2017; Reddy et al., 2021a). The lack of government extension services further complicates this situation, leaving farmers without the technical guidance needed to implement antifragile practices effectively (Anderson and Feder, 2004). In addition, the non-availability of farm machinery, coupled with inadequate irrigation infrastructure, limits the ability of farmers to enhance productivity and resilience (Hazell and Rahman, 2014).

MARKET AND INPUT CONSTRAINTS

The non-availability of key inputs, such as high-quality seeds and fertilizers, is a major hurdle for farmers. The prevalence of spurious seeds and fertilizers in the market further undermines their efforts to adopt antifragile practices (World Bank, 2007). Moreover, many farmers are located far from markets, making it difficult to access necessary inputs and sell their produce at fair prices (Reardon et al., 2003; Reddy and Mehjabeen, 2019). This geographic isolation is exacerbated by a lack of access to reliable information, limiting farmers' ability to make informed decisions about their practices (Goyal, 2010).

EDUCATIONAL AND KNOWLEDGE CONSTRAINTS

Low levels of education among farmers hinder their ability to understand and implement antifragile practices (Duflo et al., 2008). This knowledge gap is often exploited by private seed companies, which may pressure farmers into purchasing their products instead of adopting sustainable, diversified practices (Schnurr, 2019). Without proper education and access to unbiased information, farmers remain vulnerable to external pressures and are less likely to invest in practices that could enhance their resilience (Feder et al., 1985; Reddy, 2019).

ENVIRONMENTAL CONSTRAINTS

Environmental factors such as low productive soils and inadequate irrigation systems make it difficult for farmers to adopt antifragile practices that often require a baseline level of soil fertility and water availability (Pretty, 2008). In areas with degraded soils or scarce water resources, the potential benefits of antifragile practices are less apparent, further discouraging their adoption (Autio et al., 2021; Reddy et al., 2021b).

The constraints to adopting antifragile practices are multifaceted and rooted in economic, labour, market, educational, and environmental challenges. Addressing these barriers requires a holistic approach that improves access to financial resources, enhances government extension services, provides better market access, and empowers farmers through education and information. Only by tackling these interconnected constraints can farmers be supported in adopting antifragile practices that ensure long-term sustainability and resilience.

DISCUSSION ON PATHWAYS FOR FARMERS TO ADOPT ANTIFRAGILE POLICIES

Adopting antifragile policies in agriculture, which allow farmers to thrive in volatile and uncertain environments, requires a multifaceted approach. This discussion outlines the key pathways and strategies for enabling farmers to embrace such practices, drawing on influential studies and frameworks.

Education and Awareness

The role of education and awareness in fostering antifragile farming systems cannot be overstated. Taleb (2012) highlights that antifragility thrives on learning from small disruptions, and this principle can be applied to agriculture through targeted training programmes. These programmes should focus on educating farmers about how practices like crop diversification and agroforestry can not only mitigate risks but also create opportunities for growth. Pretty et al. (2011) emphasize the importance of peer learning networks, where farmers share experiences and collectively refine their practices. Agricultural extension services also play a pivotal role in disseminating knowledge on resilient farming methods, as highlighted by Anderson and Feder (2007).

Access to Resources

Resource access is another critical pathway to antifragility. Financial instruments such as microcredit and crop insurance, tailored to promote resilience, have been shown to be effective in several regions (Zeller and Sharma, 2000). Additionally, Binswanger-Mkhize (2012) argues for subsidies that specifically incentivize sustainable practices like organic farming or water conservation. Establishing local resource centres that provide inputs and technical assistance is also crucial. These centres can serve as hubs for distributing drought-resistant seeds or promoting innovative farming technologies.

Policy Support

Policy frameworks must be conducive to antifragile practices. Ellis (1993) stresses the need for policies that support smallholder farmers and enable land reforms. Regulatory flexibility is essential to allow farmers to experiment and innovate, reducing bureaucratic barriers. National and local agricultural strategies should incorporate antifragile principles, aligning with the broader goals of sustainability and resilience, as suggested by Scoones (1998). The integration of antifragility into policy ensures long-term support for farmers as they transition to more resilient systems.

Research and Development

Collaborative research between farmers, academic institutions, and agricultural bodies is vital for developing antifragile practices. Chambers and Conway (1992) advocate for participatory research approaches that include farmers in the development of solutions, ensuring that innovations are context-specific and scalable. Pilot projects can serve as testing grounds for new practices, and those that prove successful should be scaled up. Monitoring and feedback mechanisms, as proposed by Pretty (1995), are crucial for refining these practices and ensuring continuous improvement.

Market Access and Diversification

Market access is a significant factor in encouraging antifragility. Barrett (2008) notes that connecting farmers to markets, particularly for diversified and resilient products, can reduce dependency on single commodities and enhance income stability. Value addition at the farm level, such as through processing or packaging, can create new revenue streams, making farmers less vulnerable to price fluctuations. Davis and Goldberg (1957) highlight the importance of collective marketing efforts, which can strengthen farmers' bargaining power and reduce market risks.

Technology and Innovation

The adoption of technology is a key driver of antifragility. Pingali (2012) underscores the potential of smart farming technologies, such as precision agriculture, to enhance decision-making and resource use efficiency. Low-cost innovations, such as solar irrigation pumps, are particularly relevant for smallholders, as they provide scalable solutions that do not require significant capital investment. Open-source platforms, where farmers can access information and share experiences, further support the diffusion of antifragile practices, as evidenced by the work of Douthwaite et al. (2001).

Social and Community Engagement

Building social capital and engaging communities in decision-making processes are essential for sustaining antifragile practices. Ostrom (1990) emphasizes the role of collective action in managing common resources and ensuring community resilience. By involving local communities in planning and decision-making, a sense of ownership and responsibility is fostered, leading to more sustainable outcomes. Social networks also play a critical role, as they enable farmers to support one another in adopting and maintaining new practices.

Resilience-Building through Diversification

Diversification is a cornerstone of antifragile systems. Ellis (2000) argues that diversification, whether through crop variety, livestock, or income sources, reduces risks and enhances resilience. Agroforestry and mixed farming practices, which promote biodiversity and ecological balance, are examples of diversification strategies that build resilience against environmental shocks. Encouraging income diversification beyond agriculture, such as through eco-tourism or handicrafts, can further reduce vulnerability and increase antifragility.

CONCLUSION

In conclusion, adopting antifragile policies in agriculture requires a holistic approach that encompasses education, resource access, policy support, research, market integration, technology adoption, community engagement, and diversification. By following these pathways, farmers can transition to systems that not only withstand shocks but also grow stronger in response to them. Along with this, it is crucial for policymakers and technology developers to understand farmers' needs and their capacity to adopt new technologies, ensuring that the innovations are well suited to their requirements (Mwangi, 2015).

REFERENCES

Acevedo M, Pixley K, Zinyengere N, Meng S, Tufan H, Cichy K, Bizikova L, Isaacs K, Ghezzi-Kopel K, Porciello J. (2020). A Scoping Review of Adoption of Climate-Resilient Crops by Small-Scale Producers in Low-and Middle-Income Countries. *Nature Plants*, 6(10), pp. 1231–1241.

Anderson JR, Feder G. (2004). Agricultural Extension: Good Intentions and Hard Realities. *The World Bank Research Observer*, 19(1), pp. 41–60, https://doi.org/10.1093/wbro/lkh013

Anderson JR, Feder G. (2007). Agricultural Extension. *Handbook of Agricultural Economics*, 3, pp. 2343–2378.

Autio A, Johansson T, Motaroki L, Minoia P, Pellikka P. (2021). Constraints for Adopting Climate-Smart Agricultural Practices among Smallholder Farmers in Southeast Kenya. *Agricultural Systems*, 194, 103284.

Bardhan P. (1984). *Land, Labor, and Rural Poverty: Essays in Development Economics*. Columbia University Press.
Barrett CB. (2008). Smallholder Market Participation: Concepts and Evidence from Eastern and Southern Africa. *Food Policy*, 33(4), pp. 299–317. https://doi.org/10.1016/j.foodpol.2007.10.005
Barrett CB, Reardon T, Webb P. (2001). Nonfarm Income Diversification and Household Livelihood Strategies in Rural Africa: Concepts, Dynamics, and Policy Implications. *Food Policy*, 26(4), pp. 315–331.
Binswanger-Mkhize HP. (2012). Is There Too Much Hype about Index-Based Agricultural Insurance? *Journal of Development Studies*, 48(2), pp. 187–200.
Chambers R, Conway GR. (1992). *Sustainable Rural Livelihoods: Practical Concepts for the 21st Century*. IDS Discussion Paper, No. 296.
Chazovachii B, Chigwenya A, Mushuku A. (2012). Adoption of Climate Resilient Rural Livelihoods through Growing of Small Grains in Munyaradzi Communal Area, Gutu District. *African Journal of Agricultural Research*, 7(8), pp. 1335–1345.
Davis JH, Goldberg RA. (1957). *A Concept of Agribusiness*. Division of Research, Graduate School of Business Administration, Harvard University.
Douthwaite B, Keatinge JDH, Park JR. (2001). Why Promising Technologies Fail: The Neglected Role of User Innovation during Adoption. *Research Policy*, 30(5), pp. 819–836.
Duflo E, Kremer M, Robinson J. (2008). How High Are Rates of Return to Fertilizer? Evidence from Field Experiments in Kenya. *American Economic Review*, 98(2), pp. 482–488.
Ellis F. (1993). *Peasant Economics: Farm Households and Agrarian Development*. Cambridge University Press.
Ellis F. (2000). *Rural Livelihoods and Diversity in Developing Countries*. Oxford University Press.
FAO. (2017). *The Future of Food and Agriculture: Trends and Challenges*. Food and Agriculture Organization of the United Nations.
Feder G, Just RE, Zilberman D. (1985). Adoption of Agricultural Innovations in Developing Countries: A Survey. *Economic Development and Cultural Change*, 33(2), pp. 255–298.
Goyal A. (2010). Information, Direct Access to Farmers, and Rural Market Performance in Central India. *American Economic Journal: Applied Economics*, 2(3), pp. 22–45.
Hazell P, Rahman A. (2014). *New Directions for Smallholder Agriculture*. Oxford University Press.
Jain M, Naeem S, Orlove B, Modi V, DeFries RS. (2015). Understanding the Causes and Consequences of Differential Decision-Making in Adaptation Research: Adapting to a Delayed Monsoon Onset in Gujarat, India. *Global Environment Change*, 31, pp. 98–109.
Mulugeta G. (2014). Evergreen Agriculture: Agroforestry for Food Security and Climate Change Resilience. *Journal of Natural Sciences Research*, 4(11), pp. 80–90.
Mwangi M, Kariuki S. (2015). Factors Determining Adoption of New Agricultural Technology by Smallholder Farmers in Developing Countries. *Journal of Economics and sustainable development*, 6(5), pp. 208–216.
Ostrom E. (1990). *Governing the Commons: The Evolution of Institutions for Collective Action*. Cambridge University Press.
Pankaj PK, Prasad JV, Singh VK, Nirmala G, Reddy KS. (2024). Climate Resilient Livestock Farming Systems. In Ayon Tarafdar, Ashok Pandey, Gyanendra Kumar Gaur, Mukesh Singh, Hari Om Pandey (Eds.), *Engineering Applications in Livestock Production* (pp. 159–178). Academic Press.
Pingali PL. (2012). Green Revolution: Impacts, Limits, and the Path Ahead. *Proceedings of the National Academy of Sciences*, 109(31), pp. 12302–12308.
Pretty JN. (1995). *Regenerating Agriculture: Policies and Practice for Sustainability and Self-Reliance*. Earthscan, London, p. 320.
Pretty J. (2008). Agricultural Sustainability: Concepts, Principles and Evidence. *Philosophical Transactions of the Royal Society B: Biological Sciences*, 363(1491), pp. 447–465.

Pretty J, Toulmin C, Williams S. (2011). Sustainable Intensification in African Agriculture. *International Journal of Agricultural Sustainability*, 9(1), pp. 5–24. https://doi.org/10.3763/ijas.2010.0583

Reardon T, Timmer CP, Barrett CB, Berdegué J. (2003). The Rise of Supermarkets in Africa, Asia, and Latin America. *American Journal of Agricultural Economics*, 85(5), pp. 1140–1146. https://doi.org/10.1111/J.0092-5853.2003.00520.X

Reddy AA. (2012). Structure of Indebtedness of Households in Semi-Arid Tropics of India. *Agricultural Economics Research Review*, 25, pp. 473–483.

Reddy AA. (2019). The Soil Health Card Scheme in India: Lessons Learned and Challenges for Replication in Other Developing Countries. *Journal of Natural Resources Policy Research*, 9(2), pp. 124–156.

Reddy AA, Bhattacharya A, Reddy SV, Ricart S. (2021a). Farmers' Distress Index: An Approach for an Action Plan to Reduce Vulnerability in the Drylands of India. *Land*, 10(11), p. 1236.

Reddy AA, and Mehjabeen. (2019). Electronic National Agricultural Markets, Impacts, Problems and Way Forward. *IIM Kozhikode Society & Management Review*, 8(2), pp. 143–155.

Reddy AA, Singha Roy N, Pradeep D. (2021b). Has India's Employment Guarantee Program Achieved Intended Targets? *Sage Open*, 11(4), p. 21582440211052281.

Scheffran J, Marmer E, Sow P. (2012). Migration as a Contribution to Resilience and Innovation in Climate Adaptation: Social Networks and Co-development in Northwest Africa. *Applied Geography*, 33, pp. 119–127.

Schnurr MA. (2019). *Africa's Gene Revolution: Genetically Modified Crops and the Future of African Agriculture*. McGill-Queen's University Press. London, p. 336.

Scoones I. (1998). *Sustainable Rural Livelihoods: A Framework for Analysis*. IDS Working Paper, No. 72, Institute of Development Studies.

Taillandier C, Cörvers R, Stringer LC. (2023). Growing Resilient Futures: Agroforestry as a Pathway towards Climate Resilient Development for Smallholder Farmers. *Frontiers in Sustainable Food Systems*, 7, p. 1260291.

Taleb NN. (2012). *Antifragile: Things That Gain from Disorder*. Random House.

Vishnoi L, Kumar A, Kumar S, Sharma G, Baxla AK, Singh KK, Bhan SC. (2020). Weather Based Crop Insurance for Risk Management in Agriculture. *Journal of Agrometeorology*, 22(2), pp. 101–108.

World Bank. (2007). *Agriculture for Development*. World Development Report 2008. The World Bank.

Zeller M, Sharma M. (2000). Many Borrow, More Save, and All Insure: Implications for Food and Micro-Finance Policy. *Food Policy*, 25(2), pp. 143–167.

11 Understanding Farmers' Perception of Risk

After understanding the numerous struggles in farmers' lives, we felt it was very important to know how they perceive risks and make decisions about new farming practices (Douxchamps et al., 2016; Kisaka-Lwayo and Obi, 2012; Schlecht et al., 2006), as it determined how they deal with the uncertainty. In understanding farmers' risk-taking behaviours, it is crucial to consider the socio-economic factors that shape their decision-making. Studies have shown that risk-averse farmers often exhibit lower levels of adoption of new technologies and practices, as they prioritize minimizing potential losses over pursuing higher gains (Koundouri et al., 2006). This cautious approach can be linked to concerns about financial security, particularly in the context of taking loans, where fear of debt may inhibit investment in modern agricultural inputs (Binswanger, 1980). On the contrary, those willing to take risks may be more likely to embrace innovations that promise increased productivity, reflecting a broader trend towards specialization and efficiency in agricultural practices (Feder et al., 1985). This is so because, to handle inevitable periods of change, farmers need to maintain diversity and flexibility for better adaptability (Darnhofer et al., 2010). Moreover, the adoption of new technologies and crop varieties is often associated with farmers' perceptions of risk and reward. Research indicates that farmers who are less concerned about the risks associated with loans and financial instability are more inclined to invest in these innovations, viewing them as opportunities to enhance their competitiveness (Rogers, 2003). This aligns with the broader theory of technology adoption, where individuals weigh the perceived benefits against the potential risks and costs (Pannell et al., 2006). Specialization in specific crops is also a key indicator of a farmer's risk tolerance, with those focusing on a single crop often relying on advanced techniques and inputs to maximize yield and profit, despite the inherent risks (Chavas and Holt, 1996).

This will help identify options to make farmers more adaptable (Cooper et al., 2008; Milgroom and Giller, 2013). We learned that their past experiences, knowledge, culture, habits, and beliefs greatly influenced their perception and attitude (van Winsen et al., 2011). For example, if a farmer had a bad experience with a new technology before, they might be hesitant to try it again. On the contrary, farmers who have successfully adopted new practices in the past are more open-minded. Apart from these, social norms and peer pressure matter. If farmers saw others in their community embracing a new technology, they were more likely to try it too. But if there was resistance, they might avoid it. Economic factors, too, had a strong influence. Farmers carefully considered the costs and potential profits of new methods, balancing the risks. So, understanding these factors was crucial for developing helpful interventions. Every decision they made, from what crops to plant to when to harvest, involved uncertainty. One farmer shared, "*I still*

DOI: 10.1201/9781003582878-11

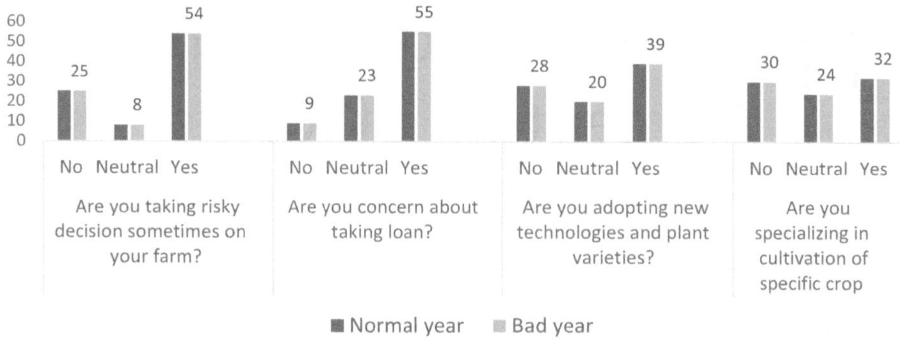

FIGURE 11.1 Risk perception by farmers (% of farmers).

remember a crop failure that hurt me deeply. It made me alert to new technologies. I hesitate to use any new things because I am afraid it would lead to more losses. Past experiences shaped our perceptions of how we see things, making us cautious about potential problems."

As we spent time listening to the farmers' stories, we started to understand their beliefs and practices. We discovered that more than half of them were true risk-takers not just in their agricultural practices but in every sense. This intrigued us, and we wanted to understand it further. So, we decided to speak to some of the farmers to understand their mindset better. One middle-aged farmer shared his feeling of depression, saying:

We are at the mercy of nature. We plant our seeds and pray for a good harvest, but sometimes, nature has other plans. We must take risks to survive.

His words resonated with us, and we began to see things from a different perspective. We figured out that about 55% of farmers were concerned about taking loans, as they feared being unable to repay them (Figure 11.1). This fear often led them to stick to traditional farming methods, even though adopting new technologies and plant varieties could potentially increase their yield. We realized that the farmers never take unnecessary risks; rather, their perceptions were shaped by their experiences and the challenges they faced. They were not afraid of taking risks; they were afraid of losing everything they had worked so hard for.

EXPLORING FARMERS' PERSPECTIVES ACROSS DIFFERENT GROUPS

Although there was a good practice of local adaptation, these may not be enough to fully counter the growing impacts of climate change (Dhanya and Ramachandran, 2016). Further, regardless of how many resources farmers have, they all face various risks. But what was interesting was that each farmer saw these risks differently (Huet et al., 2020). So, to gain a better understanding of the risk perception of farmers, we focused on various groups. We wanted to know if the financial status and access to resources really played a significant role in shaping the farmers' attitudes and behaviours.

Religious Groups

Religion somehow has some influence on farmers' risk perception and risk attitude (Ahmad et al., 2020). In a study in Ethiopia, it was found that Christians showed higher risk-taking tendencies than other religions (Feyisa et al., 2023). We found the replicate in our research. We found that 61% of Christian farmers are interested in taking loans (Figure 11.2). So it was clear that religion played a significant role in the risk-taking behaviour of these farmers. To understand the reasoning behind this, we delved deeper into our interviews. We explored that the strong belief in God of Christian farmers helps them to take risks, knowing that God would take care of them. It was a risky decision for these farmers to rely on their faith to make financial decisions, but it seemed to be paying off for them.

Social Groups

We discovered that the division in the social groups affects not only the farmers' social status but also their mindset when it comes to taking risks. Among different social groups, general category farmers were the most worried, with 65% of them expressing concerns about taking a loan. This was because they were often burdened with societal expectations and the pressure to maintain their status. One farmer reluctantly admitted:

Taking a loan is seen as a sign of failure, and it brings great shame to us.

However, the most surprising finding was that of the ST farmers, with only 36% of them expressing concerns about loans (Figure 11.3). As the most marginalized group,

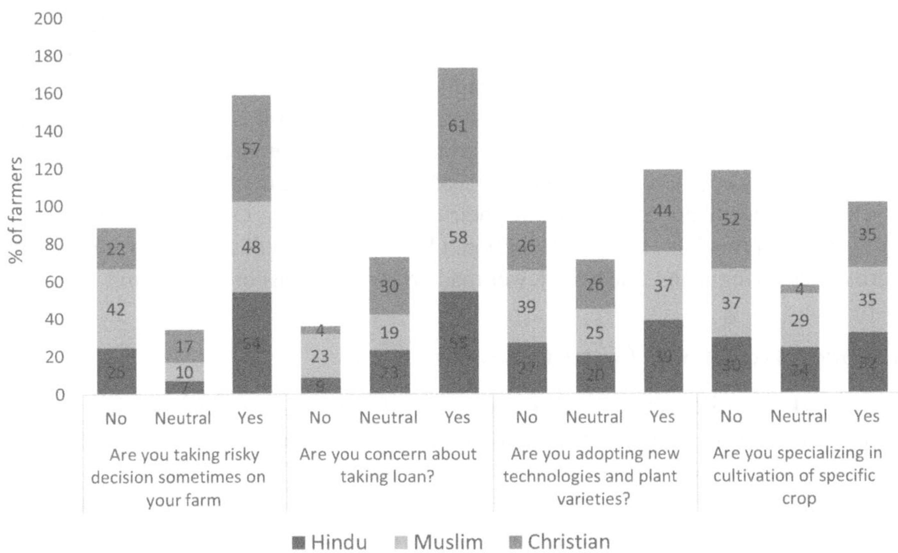

FIGURE 11.2 Decision-making under risk by religion.

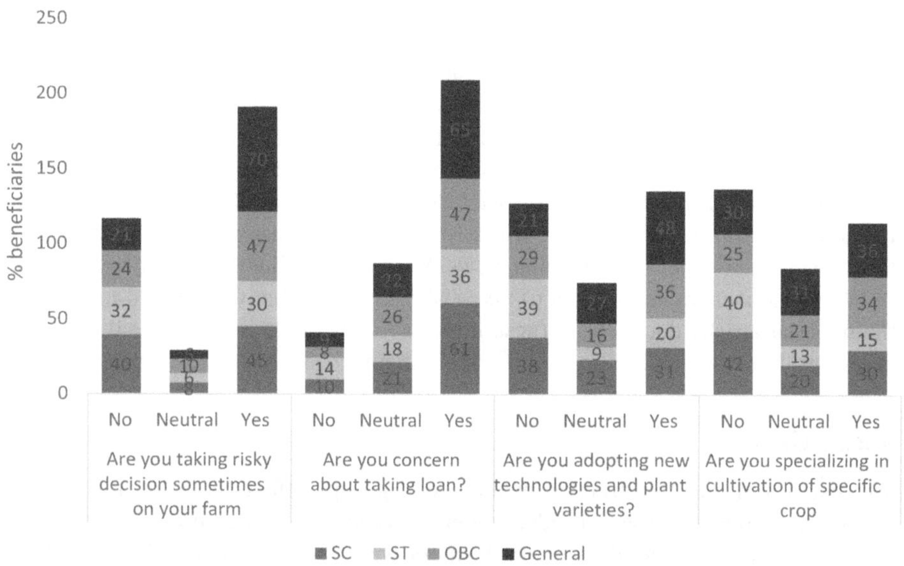

FIGURE 11.3 Decision-making under risk by social group.

they have learned to live with risk and uncertainty. They were used to making bold and risky decisions, and this was reflected in their attitude towards loans.

Landholding Groups (Own Land)

As we delved into the landholding group (own land), we discovered the differences in the attitudes and behaviours of farmers. About 58% of large and medium landholding farmers were most concerned about taking loans (Figure 11.4). It seemed that the more land a farmer had, the higher their risk perception was. But it was not just loans that were a cause for concern; being financially strong and well-connected to higher officials, large farmers were more likely to make risky decisions on their farms as it gave them a sense of security and confidence in their decision-making. On the contrary, smaller farmers had a much lower risk perception. Only 48% of small farmers expressed concern about taking loans, and even fewer made risky decisions on their farms.

Landholding Groups (Total Cultivable Land)

We found a similar trend in this group too. Larger farmers had a high level of concern regarding taking loans, with about 58% of them expressing their worries (Figure 11.5). Conversely, the smaller farmers had a low level of concern, with only 48% of them being worried about taking loans. This trend was also reflected in their interest in adopting new technologies and plant varieties. Large farmers were more open to adopting new technologies and specializing in the cultivation of specific crops, while the smaller farmers were more reluctant to do that.

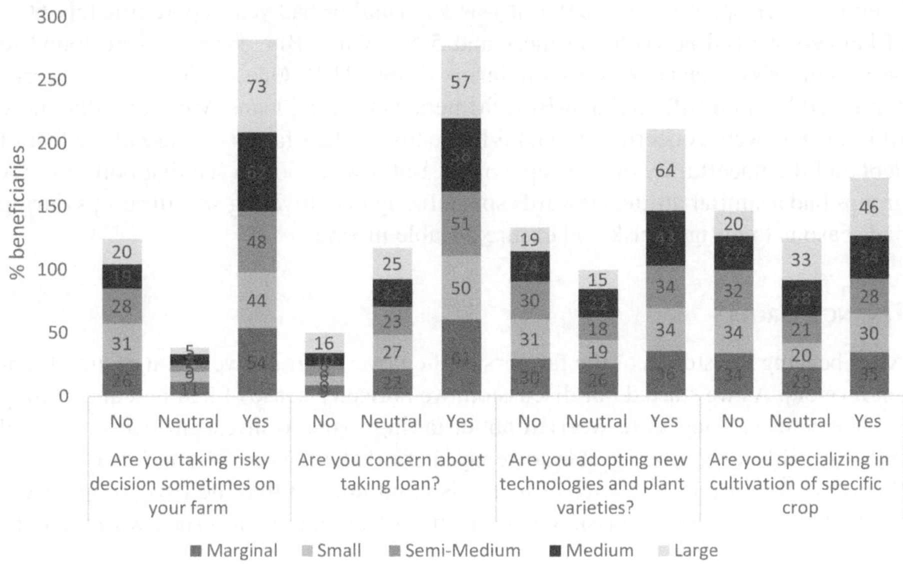

FIGURE 11.4 Decision-making under risk by land size (own).

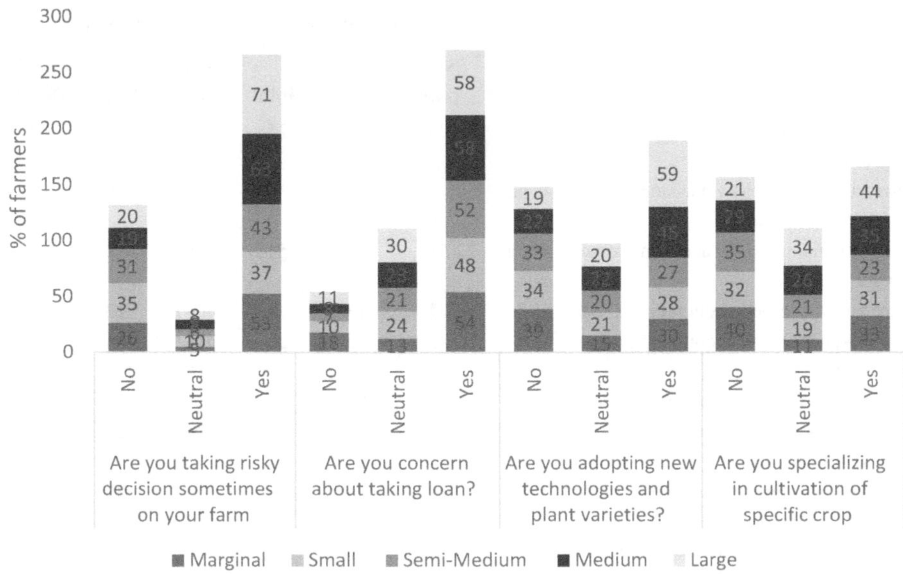

FIGURE 11.5 Decision-making under risk by land cultivated category.

Poverty Groups

We paid special attention to the risk perception of farmers belonging to different poverty groups as we knew very well that poverty can greatly affect one's perception of risk, and this is especially true for farmers who rely on their land for their livelihood. Irrespective of whether it was a normal or bad year, approximately 60% of below-poverty-line (BPL) farmers and 50% of non-BPL farmers were found to be making risky decisions on their farms (Table 11.1). One of the main concerns expressed by both BPL and non-BPL farmers was taking loans. We found that more BPL farmers were concerned about taking loans for their farms because of the fear of debt and the uncertainty of non-repayment. But it was good to see that both poverty groups had a similar attitude towards specializing in cultivating specific crops during bad seasons to mitigate risk and ensure a stable income.

Tenancy Groups

After hearing the stories of the farmers of the poverty group, we met a group of tenant farmers. As we started our discussion, we noticed a sense of fear and uncertainty in their voices. They were worried about taking loans to invest in their crops and the possibility of not being able to repay them if the harvest was not as expected. We spoke to a group of owner farmers also and noticed that the two groups had a vastly different outlook on risk. Owner farmers were more concerned with the risks

TABLE 11.1
Decision-Making under Risk by Poverty Status

	Normal Year			Bad Year		
	Non-poor	Poor	Total	Non-poor	Poor	Total
Are you making risky decisions sometimes on your farm? No	25	26	25	30	20	25
Are you making risky decisions sometimes on your farm? Neutral	7	8	8	8	8	8
Are you making risky decisions sometimes on your farm? Yes	51	61	54	51	58	54
Are you concerned about taking a loan? No	7	13	9	9	10	9
Are you concerned about taking a loan? Neutral	22	25	23	23	23	23
Are you concerned about taking a loan? Yes	54	56	55	57	52	55
Are you adopting new technologies and plant varieties? No	26	31	28	29	27	28
Are you adopting new technologies and plant varieties? Neutral	19	24	20	20	21	20
Are you adopting new technologies and plant varieties? Yes	39	39	39	41	37	39
Are you specializing in the cultivation of a specific crop? No	25	41	30	29	32	30
Are you specializing in the cultivation of a specific crop? Neutral	26	20	24	27	21	24
Are you specializing in the cultivation of a specific crop? Yes	32	32	32	33	32	32

involved in making decisions regarding their farms because they had more to lose if their crops failed (Figure 11.6). But interestingly both groups had similar levels of risk perception when it came to specialized crops and adopting new technologies. This meant that the fear of failure was universal among farmers, regardless of their status as tenants or owners.

Irrigation Groups

In the irrigation group, we also observed a vast difference between dryland and irrigated farmers when it came to their attitudes towards loans and risk-taking. Dryland farmers were much more worried about taking loans than irrigated farmers. But it did not end there. We saw that irrigated farmers with access to water were more prone to making risky decisions on their farms (Figure 11.7). This was a revelation to us as we always thought that irrigated farmers had it easy, but, in reality, they were just as challenged as us in avoiding risky decision-making.

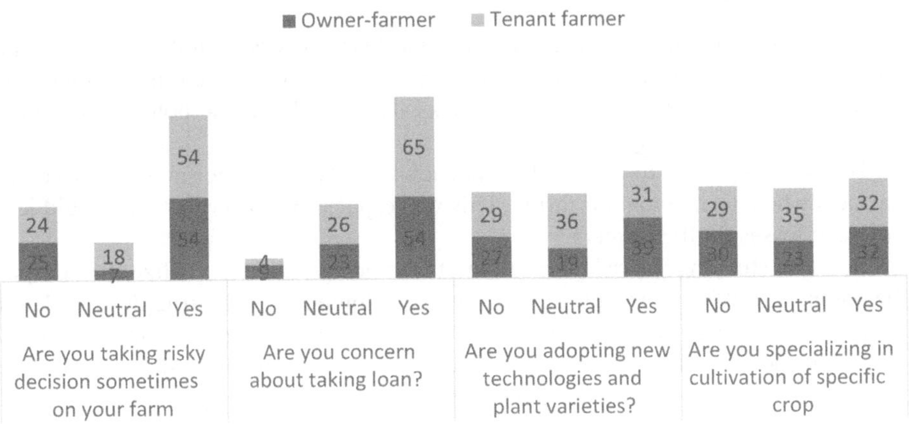

FIGURE 11.6 Decision-making under risk by tenant farmers (% of farmers).

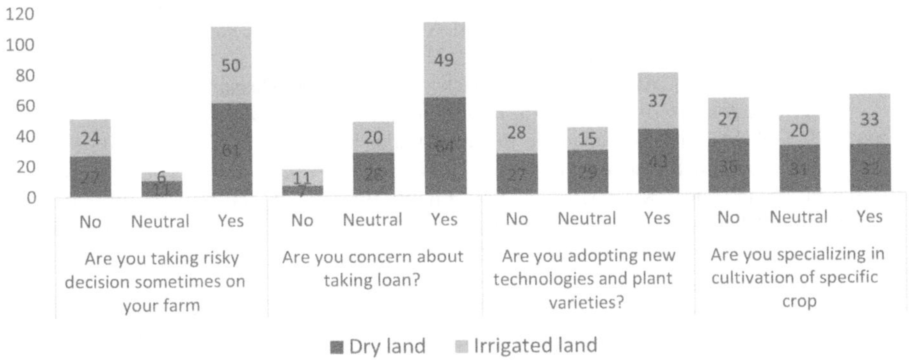

FIGURE 11.7 Decision-making under risk by dryland and irrigated farmers (% of farmers).

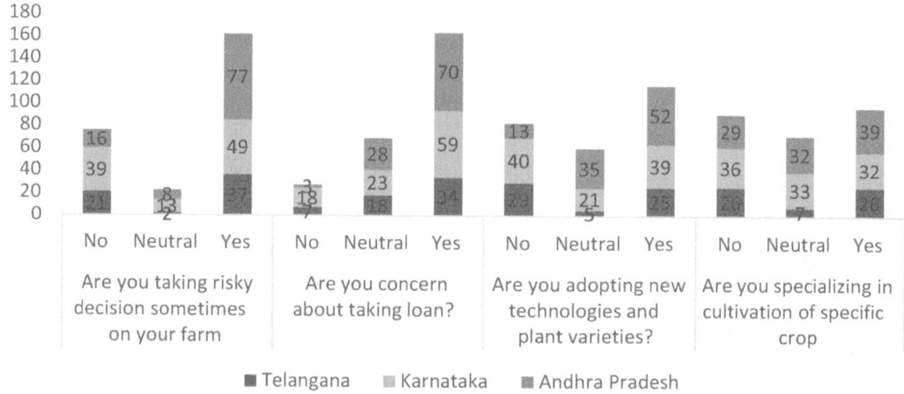

FIGURE 11.8 Decision-making under risk by state (% of farmers).

Comparison among States

When we met the farmers we could sense the fear and worry in their voices in all three states. We noticed that the farmers in Andhra Pradesh were most concerned about taking loans, as they saw it as a risky decision that could potentially lead to financial ruin (Figure 11.8).

But what stood out to us was the tendency to make risky decisions of Karnataka farmers when it came to their farms. This was especially concerning as it could lead to major losses for the farmers.

So, after studying the risk perception of farmers, we came to realize that, despite belonging to different groups, farmers were all united in their struggles and fears. Regardless of their background, they shared the common goal of making a living from the land. The burden of risk perception was something that bound them together, reminding them of the shared challenges they face in agriculture.

REFERENCES

Ahmad D, Afzal M, Rauf A. (2020). Environmental Risks among Rice Farmers and Factors Influencing Their Risk Perceptions and Attitudes in Punjab, Pakistan. *Environmental Science and Pollution Research*, 27, pp. 21953–21964.

Binswanger HP. (1980). Attitudes toward Risk: Experimental Measurement in Rural India. *American Journal of Agricultural Economics*, 62(3), pp. 395–407.

Chavas JP, Holt MT. (1996). Economic Behavior under Uncertainty: A Joint Analysis of Risk Preferences and Technology. *The Review of Economics and Statistics*, 78(2), pp. 329–335.

Cooper PJ, Dimes J, Rao KP, Shapiro B, Shiferaw B, Twomlow S. (2008). Coping Better with Current Climatic Variability in the Rain-Fed Farming Systems of Sub-Saharan Africa: An Essential First Step in Adapting to Future Climate Change. *Agriculture, Ecosystems & Environment*, 126(1–2), pp. 24–35.

Darnhofer I, Fairweather J, Moller H. (2010). Assessing a Farm's Sustainability: Insights from Resilience Thinking. *International Journal of Agricultural Sustainability*, 8(3), pp. 186–198.

Dhanya P, Ramachandran A. (2016). Farmers' Perceptions of Climate Change and the Proposed Agriculture Adaptation Strategies in a Semi Arid Region of South India. *Journal of Integrative Environmental Sciences*, *13*(1), pp. 1–18.

Douxchamps S, Van Wijk MT, Silvestri S, Moussa AS, Quiros C, Ndour NY, Buah S, Somé L, Herrero M, Kristjanson P, Ouedraogo M. Philip KT, Piet VA, Robert Z, MC Rufino (2016). Linking Agricultural Adaptation Strategies, Food Security and Vulnerability: Evidence from West Africa. *Regional Environmental Change*, 16, pp. 1305–1317.

Feder G, Just RE, Zilberman D. (1985). Adoption of Agricultural Innovations in Developing Countries: A Survey. *Economic Development and Cultural Change*, 33(2), pp. 255–298.

Feyisa AD, Maertens M, de Mey Y. (2023). Relating Risk Preferences and Risk Perceptions over Different Agricultural Risk Domains: Insights from Ethiopia. *World Development*, 162, p. 106137.

Huet EK, Adam M, Giller KE, Descheemaeker K. (2020). Diversity in Perception And Management of Farming Risks in Southern Mali. *Agricultural Systems*, 184, Article no. 102905. https://doi.org/10.1016/j.agsy.2020.102905

Kisaka-Lwayo M, Obi A. (2012). Risk Perceptions and Management Strategies by Smallholder Farmers in KwaZulu-Natal Province, South Africa. *International Journal of Agricultural Management*, 1(3), pp. 28–39.

Koundouri P, Nauges C, Tzouvelekas V. (2006). Technology Adoption under Production Uncertainty: Theory and Application to Irrigation Technology. *American Journal of Agricultural Economics*, 88(3), pp. 657–670.

Milgroom J, Giller KE. (2013). Courting the Rain: Rethinking Seasonality and Adaptation to Recurrent Drought in Semi-Arid Southern Africa. *Agricultural Systems*, 118, pp. 91–104.

Pannell DJ, Marshall GR, Barr N, Curtis A, Vanclay F, Wilkinson R. (2006). Understanding and Promoting Adoption of Conservation Practices by Rural Landholders. *Australian Journal of Experimental Agriculture*, 46(11), pp. 1407–1424.

Rogers EM. (2003). *Diffusion of Innovations* (5th ed.). Free Press. New Yark.

Schlecht E, Buerkert A, Tielkes E, Bationo A. (2006). A Critical Analysis of Challenges and Opportunities for Soil Fertility Restoration in Sudano-Sahelian West Africa. *Nutrient Cycling in Agroecosystems*, 76, pp. 109–136.

van Winsen F, Wauters E, Lauwers L, De Mey Y, Van Passel S, Vancauteren M. (2011). Combining Risk Perception and Risk Attitude: A Comprehensive Individual Risk Behaviour Model. In *EAAE 2011 Congress*, Zürich, Switzerland, 1 January 2011, pp. 1–12.

12 Psychological Stress
Another Burden of Farmers

We have seen that the farmers' past experiences and struggles with new technology made them hesitant to adopt new things. Further, recent climatic disasters like floods, storms, and drought have caused repeated crop failures, pushing agriculture into a severe crisis. We were already aware of the struggles that come with farming, but there was another struggle that was often overlooked – the psychological stress of farmers. This overall situation has led to immense psychological stress for the farmers, resulting in numerous cases of suicide (Pandey, 2022).

This study introduced us to the epidemic of farmer suicide in India due to excessive mental stress. We met an old woman and her daughter-in-law and heard their tragic story. Silent tears flowed down the young woman's cheek. A year before our visit, the young woman's husband who was the bread earner of the family committed suicide in despair. He drank a bottle of pesticide in the field that nourished his dreams. As crops failed and debt mounted, the overbearing weight of despair crushed his will to fight, and his choice left behind his family.

Their story made us aware that the burden of being a farmer was imposed on them both physically and psychologically. When the disaster came, they worked tirelessly, day and night, trying to restore what they could. The stress and pressure of trying to keep their farm alive weighed heavily on their mind. They started to feel a sense of hopelessness and despair. So this psychological condition became a major concern (Chiswell, 2023) as it also influences their risk perception. When we took a look into our secondary research, we were astonished at how deeply the psychological conditions of farmers related to agricultural distress (Vasavi, 1999; Deshpande, 2002; Uutela, 2010; Merriott, 2016). Farmers who accepted stress and had low anxiety were less distressed (Gunn et al., 2021). To get a better insight into this issue, we spoke to the farmers and we realized that it was not just about the natural disaster. They said that the constant financial stress and thus the feelings of inability to fulfil the family's responsibility properly, along with the lack of moral support, make them psychologically upset. Also, the mental harassment due to the loan repayment or some serious conflicts with the society made a constant burden to the farmers, which never seems to go away. As a result, they lost their self-confidence and made both agriculture and their life vulnerable.

The psychological distress experienced by farmers is a complex issue shaped by multiple socio-economic and personal factors. In rural communities, isolation – whether felt within the family or society – has been identified as a key contributor to mental health challenges. Farmers often report feelings of being disconnected, exacerbated by the pressures of farming life and social expectations (Alston, 2012). The burden of loan repayment adds significant strain, contributing to heightened anxiety, insomnia, and an inability to concentrate on daily tasks (Ong et al., 2019). This financial

DOI: 10.1201/9781003582878-12

pressure can also lead to conflicts within the community and families, as disputes over economic responsibilities and deteriorating financial situations arise (Conger, 1994).

The inability to fulfil family responsibilities often leaves farmers feeling inadequate and unsupported. The lack of emotional or practical support from friends and relatives can further isolate individuals, contributing to a deep sense of despair (Merriott, 2016). Financial instability, particularly the deterioration of economic status, is another critical factor that compounds stress and anxiety among farmers. This is often linked to an increase in unhealthy coping mechanisms such as alcohol and drug abuse, gambling, and other risky behaviours (Patel et al., 2016). These issues, combined with insomnia and constant pressure, impair decision-making and the ability to effectively manage farm operations (Hagen et al., 2019). Ultimately, the cumulative effect of these factors leaves many farmers feeling trapped in a cycle of stress and financial insecurity, severely impacting their mental well-being.

HOW DO PSYCHOLOGICAL STRUGGLES DIFFER AMONG VARIOUS FARMER GROUPS?

Religious Groups

We saw that the psychological issue was not just a struggle for one religion but for more than 50% of farmers regardless of their religion. When the crops failed, irrespective of religion, farmers faced the same struggle, as their prayers and beliefs were not enough to overcome the harsh reality of their financial stress. However, Christians suffer from this problem a little more (Table 12.1). The matters worsened when they did not get support from their family, friends, and relatives. Arguments and disputes over money became a regular occurrence, affecting many farmers in all communities. This added to the farmers' stress and made it difficult for them to concentrate on their work and make important decisions.

Social Groups

However, we were shocked to find that their caste significantly affected their mental well-being. Over 81% of SC farmers faced immense pressure regarding loan repayment, the highest among social groups (Table 12.2). Meanwhile, pressure on higher castes, especially OBCs, was slightly lower. Upon further investigation, we discovered that the SC and general caste farmers felt the pressure of societal expectations and discrimination, adding to their already stressful financial situation. On the contrary, although ST and OBC farmers faced similar struggles, they had stronger community support that gave them some relief.

Landholding Groups (Own Land)

Similarly, in the landholding group, the marginal farmers, the most vulnerable community, faced the highest pressure for loan repayment (Table 12.3). And the situation only got worse in 2021, with a slight increase in pressure for all categories of farmers.

TABLE 12.1
Psychological Issues Faced by Religion of Farmers (% of farmers who said "Yes")

	Religion			
	Hindu	Muslim	Christian	Total
Do you feel yourself isolated in the society or in your family? (2020)	12	15	39	12
Do you feel yourself isolated in the society or in your family? (2021)	12	15	39	13
Pressure for loan repayment (2020)	74	58	83	74
Pressure for loan repayment (2021)	76	58	83	76
Quarrel among your society (2020)	13	17	35	13
Quarrel among your society (2021)	13	17	35	14
Not able to fulfil your family's responsibilities (2020)	49	56	65	50
Not able to fulfil your family's responsibilities (2021)	52	54	65	53
Not getting support from your family/friends/relatives? (2020)	23	29	35	24
Not getting support from your family/friends/relatives? (2021)	26	29	35	26
Are you terribly worried about your financial situation? (2020)	67	62	65	66
Are you terribly worried about your financial situation? (2021)	70	64	65	70
Problem with the family members regarding the deterioration in economic status? (2020)	26	25	44	26
Problem with the family members regarding the deterioration in economic status? (2021)	29	31	44	29
Alcohol/drug abuse/betting/gambling/insomnia (2020)	11	8	4	11
Alcohol/drug abuse/betting/gambling/insomnia (2021)	12	4	4	11
Unable to concentrate on your work and make decisions (2020)	30	33	61	30
Unable to concentrate on your work and make decisions (2021)	30	33	61	31
Do you think that you have been under constant pressure for a long time? (2020)	53	62	83	54
Do you think that you have been under constant pressure for a long time? (2021)	54	62	83	55

The most distressing fact was that the farmers, particularly the smaller ones, always felt like they were in a constant state of pressure.

Landholding Groups (Total Cultivated Land)

However, in this group, we found a slightly different trend. Mostly the medium and large farmers not only faced the highest pressure for loan repayment but also felt like they were constantly under pressure for a long time (Table 12.4). This was a worrying

TABLE 12.2
Psychological Issues Faced by Social Groups (% of farmers who said "yes")

	Caste				
	SC	ST	OBC	General	Total
Do you feel yourself isolated in the society or in your family? (2020)	12	6	8	18	12
Do you feel yourself isolated in the society or in your family? (2021)	12	7	9	18	13
Pressure for loan repayment (2020)	82	72.6	71	75	74
Pressure for loan repayment (2021)	82	73.3	75	76	76
Quarrel among your society (2020)	11	8	10	18	13
Quarrel among your society (2021)	13	10	11	18	14
Not able to fulfil your family's responsibilities (2020)	53	52	50	47	50
Not able to fulfil your family's responsibilities (2021)	56	57	54	49	53
Not getting support from your family/friends/ relatives? (2020)	21	25	24	24	24
Not getting support from your family/friends/ relatives? (2021)	21	29	27	26	26
Are you terribly worried about your financial situation? (2020)	69	69	63	69	66
Are you terribly worried about your financial situation? (2021)	70	75	67	71	70
Problem with the family members regarding the deterioration in economic status? (2020)	26	20	25	29	26
Problem with the family members regarding the deterioration in economic status? (2021)	28	22	30	30	29
Alcohol/drug abuse/betting/gambling/insomnia (2020)	8	19	12	8	11
Alcohol/drug abuse/betting/gambling/insomnia (2021)	9	21	13	8	11
Unable to concentrate on your work and make decisions (2020)	34	27	25	36	30
Unable to concentrate on your work and make decisions (2021)	35	27	26	36	31
Do you think that you have been under constant pressure for a long time? (2020)	64	49	40	66	54
Do you think that you have been under constant pressure for a long time? (2021)	65	50	41	67	55

trend, and it only seemed to increase in the next year. Many farmers especially large and medium were constantly worried about their financial situation, which had also increased among all categories of farmers in 2021.

Poverty Groups

We were aware of the fact that poverty can be a determinant and result of poor mental health (Kumar and Behmani, 2017). And after having a long conversation with the farmers we also had a realization that poverty knows no boundaries.

TABLE 12.3
Psychological Issues Faced by Own-Land Size Category

	Total Own Land group				
	Marginal	Small	Semi-Medium	Medium	Large
Do you feel yourself isolated in the society or in your family? (2020)	14	15	13	11	7
Do you feel yourself isolated in the society or in your family? (2021)	14	15	14	11	7
Pressure for loan repayment (2020)	78	71	69	77	75
Pressure for loan repayment (2021)	78	73	72	79	77
Quarrel among your society (2020)	15	17	15	10	7
Quarrel among your society (2021)	17	17	16	10	8
Not able to fulfil your family's responsibilities (2020)	51	53	48	50	41
Not able to fulfil your family's responsibilities (2021)	53	56	52	53	44
Not getting support from your family/friends/relatives? (2020)	25	28	25	22	13
Not getting support from your family/friends/relatives? (2021)	27	29	27	26	18
Are you terribly worried about your financial situation? (2020)	67	69	65	66	66
Are you terribly worried about your financial situation? (2021)	70	72	69	67	74
Problem with the family members regarding the deterioration in economic status? (2020)	28	28	28	25	22
Problem with the family members regarding the deterioration in economic status? (2021)	30	28	30	28	30
Alcohol/drug abuse/betting/gambling/insomnia (2020)	6	12	12	10	13
Alcohol/drug abuse/betting/gambling/insomnia (2021)	7	14	12	11	9
Unable to concentrate on your work and make decisions (2020)	34	38	30	28	16
Unable to concentrate on your work and make decisions (2021)	34	38	30	29	19
Do you think that you have been under constant pressure for a long time? (2020)	56	55	50	57	49
Do you think that you have been under constant pressure for a long time? (2021)	57	57	51	58	49

TABLE 12.4
Psychological Issues Faced by Cultivated Land Size Category (% of farmers who said "Yes")

	Total Cultivated Land Group				
	Marginal	Small	Semi-Medium	Medium	Large
Do you feel yourself isolated in the society or in your family? (2020)	10	12	13	13	12
Do you feel yourself isolated in the society or in your family? (2021)	10	12	14	14	11
Pressure for loan repayment (2020)	70	68	70	77	77
Pressure for loan repayment (2021)	70	70	72	80	79
Quarrel among your society (2020)	10	13	15	12	12
Quarrel among your society (2021)	10	14	16	13	13
Not able to fulfil your family's responsibilities (2020)	40	47	50	52	48
Not able to fulfil your family's responsibilities (2021)	41	51	54	55	50
Not getting support from your family/friends/relatives? (2020)	24	25	26	23	19
Not getting support from your family/friends/relatives? (2021)	24	26	28	25	25
Are you terribly worried about your financial situation? (2020)	60	66	66	67	69
Are you terribly worried about your financial situation? (2021)	60	70	69	69	75
Problem with the family members regarding the deterioration in economic status? (2020)	20	22	27	26	32
Problem with the family members regarding the deterioration in economic status? (2021)	21	23	28	30	35
Alcohol/drug abuse/ betting/gambling/ insomnia (2020)	4	16	12	10	10
Alcohol/drug abuse/ betting/gambling/ insomnia (2021)	4	18	13	10	8
Unable to concentrate on your work and make decisions (2020)	35	36	31	30	23
Unable to concentrate on your work and make decisions (2021)	36	35	32	31	25
Do you think that you have been under constant pressure for a long time? (2020)	44	53	50	57	55
Do you think that you have been under constant pressure for a long time? (2021)	45	55	51	59	55

TABLE 12.5
Psychological Issues Faced by Poverty Status (% of farmers who said "Yes")

	Poverty Status					
	2021			2022		
	Non-poor	Poor	Total	Non-poor	Poor	Total
Do you feel yourself isolated in the society or in your family? (2020)	7	23	12	7	18	12
Do you feel yourself isolated in the society or in your family? (2021)	7	24	13	8	19	13
Pressure for loan repayment (2020)	72	77	74	74	73	74
Pressure for loan repayment (2021)	75	77	76	75	76	76
Quarrel among your society (2020)	7	25	13	9	18	13
Quarrel among your society (2021)	9	25	14	9	19	14
Not able to fulfil your family's responsibilities (2020)	49	50	50	49	51	50
Not able to fulfil your family's responsibilities (2021)	54	51	53	53	52	53
Not getting support from your family/friends/relatives? (2020)	21	29	24	20	29	24
Not getting support from your family/friends/relatives? (2021)	24	30	26	23	30	26
Are you terribly worried about your financial situation? (2020)	67	66	66	67	66	66
Are you terribly worried about your financial situation? (2021)	71	67	70	70	70	70
Problem with the family members regarding the deterioration in economic status? (2020)	23	33	26	23	30	26
Problem with the family members regarding the deterioration in economic status? (2021)	26	35	29	26	33	29
Alcohol/drug abuse/betting/gambling/insomnia (2020)	11	10	11	13	8	11
Alcohol/drug abuse/betting/gambling/insomnia (2021)	11	11	11	13	9	11
Unable to concentrate on your work and make decisions (2020)	27	38	30	26	36	30
Unable to concentrate on your work and make decisions (2021)	27	38	31	26	36	31
Do you think that you have been under constant pressure for a long time? (2020)	49	63	54	52	56	54
Do you think that you have been under constant pressure for a long time? (2021)	50	64	55	53	57	55

It affects people from all walks of life, and the struggles faced by the above-poverty-line (APL) and below-poverty-line (BPL) farmers were proof of that. Although some researchers already discovered that APL farmers were more obsessive and compulsive and more prone to phobia and somatic anxiety (Kumar and Behmani, 2017), we found something different. Though the BPL farmers faced the highest pressure for loan repayment, there was not much difference among the farmers of both categories who encountered financial worries (Table 12.5). Despite the significant relation between high socio-economic status and mental issues (Akhtar et al., 1975), both were equally fighting against the constant pressure they were facing for a long time.

Tenancy Groups

As we met the tenant farmers, they told us that they feel more pressure for loan repayment because they rely on landlords who charge high rents, leaving them with little profit.

This, coupled with the uncertainty of weather and crop yield, created a never-ending cycle of anxiety and worry for them. On the contrary, the owner farmers who were in a better psychological condition were also burdened with the responsibility of managing their own land, managing finances, and meeting their family's needs (Figure 12.1). This constant pressure made them fear failure.

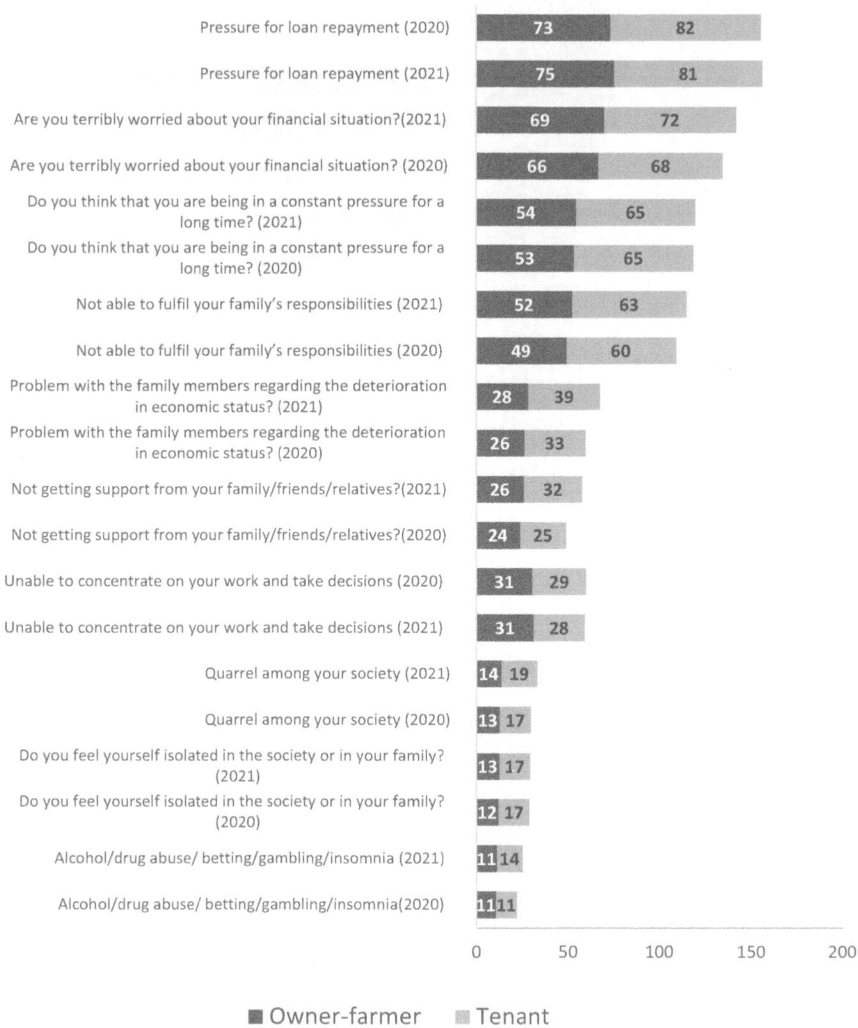

FIGURE 12.1 Psychological issues faced by tenancy status (% of farmers who said "Yes").

Irrigation Groups

As we studied dryland farmers and listened to them, we found a common story of struggle. Every day, they faced more pressure and financial worries than irrigated farmers (Figure 12.2). The burden was tremendously heavy during crop failures, with increased pressure for loan repayment. When crops fail, the pressure to repay loans has become sky-scrapping. Water scarcity and financial difficulties exacerbate their psychological distress (Yazd et al., 2020). Thus, farming became a constant battle, imposing excessive pressure on their mental health. Some farmers confessed to feeling jealous of those with irrigation facilities, "*They don't have to face the same struggles as us.*" But it was not exactly like that. Irrigated farmers were also struggling, but their struggle might be a bit easier. We realized how tough life can be for these farmers.

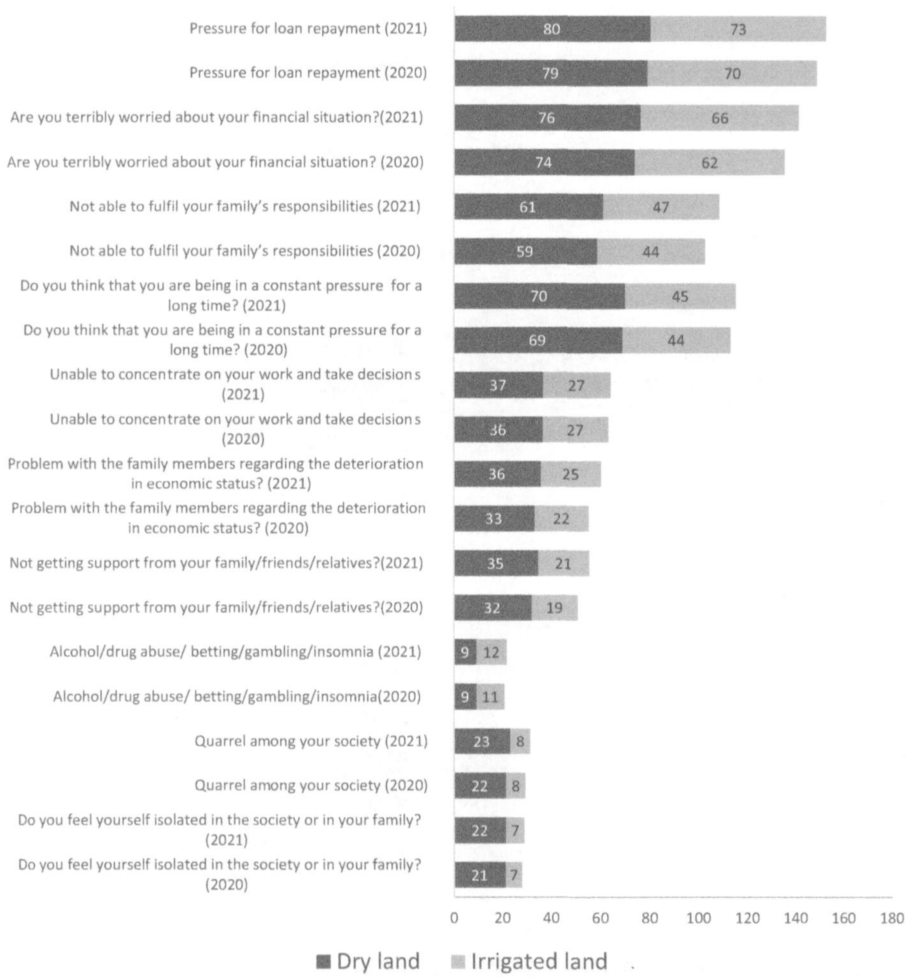

FIGURE 12.2 Psychological issues faced by irrigation status (% of farmers who said "Yes").

COMPARISON AMONG STATES

We saw that the farmers in Andhra Pradesh faced the highest pressure for loan repayment, and it got worse in 2021 (Table 12.6). It was a major reason causing the highest psychological stress among the Andhra Pradesh farmers. We realized that the root cause of their distress was totally unpredictable and was often harsh weather

TABLE 12.6
Psychological Issues Faced by Farmers in Different States (% of farmers who said "Yes")

	State Name			
	Telangana	Karnataka	Andhra Pradesh	Total
Pressure for loan repayment (2021)	64	75	87	76
Pressure for loan repayment (2020)	59	74	87	74
Are you terribly worried about your financial situation? (2021)	60	67	82	70
Are you terribly worried about your financial situation? (2020)	56	62	81	66
Do you think that you have been under constant pressure for a long time? (2021)	23	62	78	55
Do you think that you have been under constant pressure for a long time? (2020)	22	60	77	54
Not able to fulfil your family's responsibilities (2021)	49	51	58	53
Not able to fulfil your family's responsibilities (2020)	44	48	57	50
Unable to concentrate on your work and make decisions (2021)	30	18	45	31
Unable to concentrate on your work and make decisions (2020)	28	17	46	30
Problem with the family members regarding the deterioration in economic status? (2021)	19	32	36	29
Problem with the family members regarding the deterioration in economic status? (2020)	18	25	36	26
Not getting support from your family/friends/relatives? (2021)	23	25	30	26
Not getting support from your family/friends/relatives? (2020)	21	21	29	24
Quarrel among your society (2021)	7	7	27	14
Quarrel among your society (2020)	6	6	27	13
Do you feel yourself isolated in the society or in your family? (2021)	5	5	27	13
Do you feel yourself isolated in the society or in your family? (2020)	5	4	28	12
Alcohol/drug abuse/betting/gambling/insomnia(2020)	12	16	4	11
Alcohol/drug abuse/betting/gambling/insomnia (2021)	15	15	4	11

conditions that affected their crops and livelihoods. With changing climate patterns, these farmers were facing increased financial burdens, leading to a vicious cycle of stress, anxiety, and depression.

So we realized that the psychological stress was not just limited to financial pressures. The constant battle against unpredictable weather, crop failures, and market fluctuations also compelled farmers' mental well-being. It was disheartening for us to see the psychological battles that farming was taking on these hardworking individuals (Reddy et al., 2021). But, despite the challenges, they never gave up; they continued to work hard, hoping for a better future for their families.

REFERENCES

Akhtar S, Wig NN, Varma VK, Pershad D, Verma SK. (1975). A Phenomenological Analysis of Symptoms in Obsessive-Compulsive Neurosis. *The British Journal of Psychiatry*, 127(4), pp. 342–348.

Alston M. (2012). Rural Male Suicide in Australia. *Social Science & Medicine*, 74(4), pp. 515–522.

Chiswell H. (2023). Psychological Morbidity in the Farming Community: A Literature Review. *Journal of Agromedicine*, 28 (2), pp. 151–176.

Conger R. (1994). *Families in Troubled Times: Adapting to Change in Rural America* (1st ed.). Routledge. https://doi.org/10.4324/9781003058809

Deshpande RS. (2002). Suicide by Farmers in Karnataka: Agrarian Distress and Possible Alleviatory Steps. *Economic and Political Weekly*, 37(26), pp. 2601–2610.

Gunn KM, Turnbull DA, Dollman J, Kettler L, Bamford L, Vincent AD. (2021). Why Are Some Drought-Affected Farmers Less Distressed Than Others? The Association between Stress, Psychological Distress, Acceptance, Behavioural Disengagement and Neuroticism. *Australian Journal of Rural Health*, 29(1), pp. 106–116.

Hagen BN, Albright A, Sargeant J, Winder CB, Harper SL, O'Sullivan TL, Jones-Bitton A. (2019). Research Trends in Farmers' Mental Health: A Scoping Review of Mental Health Outcomes and Interventions among Farming Populations Worldwide. *PLoS One*, 14(12), e0225661.

Kumar A, Behmani RK. (2017). Effects of Poverty on Mental Health and Resilience among Rural Females of Haryana. *International Journal of Indian Psychology*, 4, pp. 25–38.

Merriott D. (2016). Factors Associated with the Farmer Suicide Crisis in India. *Journal of Epidemiology and Global Health*, 6(4), pp. 217–227.

Ong Q, Theseira W, Ng IY. (2019). Reducing Debt Improves Psychological Functioning and Changes Decision-Making in the Poor. *Proceedings of the National Academy of Sciences*, 116(15), pp. 7244–7249.

Pandey M. (2022). Exploring the Solastalgia: Predicting the Cognitive Failures among Drought Prone Farmers. *Era's Journal of Medical Research*, 9(1), pp. 39–44.

Patel V, Flisher AJ, Hetrick S, McGorry P. (2016). Mental Health of Young People: A Global Public-Health Challenge. *The Lancet*, 369(9569), pp. 1302–1313.

Reddy AA, Bhattacharya A, Reddy SV, Ricart S. (2021). Farmers' Distress Index: An Approach for an Action Plan to Reduce Vulnerability in the Drylands of India. *Land*, 10(11), p. 1236.

Uutela A. (2010). Economic Crisis and Mental Health. *Current Opinion in Psychiatry*, 23(2), pp. 127–130.

Vasavi AR. (1999). Agrarian Distress in Bidar: Market, State and Suicides. *Economic and Political Weekly*, 34(32), pp. 2263–2268.

Yazd SD, Wheeler SA, Zuo A. (2020). Understanding the Impacts of Water Scarcity and Socio-Economic Demographics on Farmer Mental Health in the Murray-Darling Basin. *Ecological Economics*, 169, p. 106564.

13 Crop Yield Losses and Economic Costs

We spend our days pouring over collecting information, analysing trends and patterns, and conducting interviews with farmers in different villages. Throughout this time spent in the world of farming, we realized how complex and challenging it is. It takes immense hard work, dedication, and resilience to succeed in this industry. We realized that there was more to farming than just planting and harvesting. It was also a business that is crucial for our society and economy. The cost of production, profits, crop loss, and insurance claims are all important factors that affect the financial stability of farmers and agricultural businesses. As we researched these factors, we began to understand the challenges that farmers face daily. Throughout this survey work, we discovered that one of the most important aspects of farming was understanding the expenses involved. From labour to equipment to seeds, every cost must be calculated to ensure maximum profit. But what happens when the prices for the farmer's products stay the same while the costs of everything else keep going up? These challenges not only hurt their profits but also cause a lot of stress. The low profits and huge losses from farming have caused a lot of distress in agriculture (NCF, 2006; Narayanamoorthy, 2013; Sainath, 2010). Some farmers have even felt so hopeless that they have taken their own lives (Narayanamoorthy, 2006, 2007; Deshpande and Arora, 2010; Huet et al., 2020).

Economic losses due to crop failure are a significant factor contributing to farmers' psychological distress. Estimates suggest that global crop losses from factors such as drought, floods, pests, and diseases can result in annual losses of up to 20–40% of total agricultural production (Savary et al., 2019). In India, for instance, a study by Reddy et al. (2020) found that recurrent droughts and erratic rainfall patterns have led to crop losses valued at billions of dollars, exacerbating rural indebtedness and contributing to a growing mental health crisis among farmers. These losses not only threaten immediate livelihoods but also deepen long-term financial instability, pushing farmers into a cycle of debt and distress (Ridley et al., 2020). Furthermore, farmers who experience consecutive crop failures often face pressure to repay loans, leading to heightened anxiety, depression, and, in extreme cases, suicide (Mishra, 2008). The economic burden of such losses is magnified in regions where there are inadequate social safety nets or crop insurance programmes, leaving farmers more vulnerable to financial ruin and psychological distress.

We can remember interviewing a farmer in Andhra Pradesh who lost his entire crop due to a sudden hailstorm. He had no insurance and was left with no way to recover his losses. It was really sad, but we heard similar stories from other farmers too. So it was a constant battle for farmers on how to maximize their profit. With narrow profit margins, every decision they make can have a significant impact on their

DOI: 10.1201/9781003582878-13

bottom line. From choosing the right seeds and fertilizers to effectively managing labour and equipment, farmers must carefully consider every aspect of their business.

Sukhdev Kaur and his family of seven, including three children, have faced crop loss multiple times before. Kaur recalled, "*Once, rains, pests and diseases and hailstorms in late March left our crops and homes destroyed in their village. It was continued over three days. My son Baljinder borrowed Rs. 6.5 lakh, with the crop ruined; there was no way to repay the loan or support the family.*" "*The hailstorm damaged the ripening crop, and then the rain waterlogged the field, causing the crops to rot,*" Baljinder added.

But even with careful planning and hard work, crop loss can occur. Weather events, pests, and diseases can all lead to devastating losses for farmers. That was the reason that insurance was crucial for protecting their livelihoods. Because, without help, these losses could be catastrophic and threaten a farmer's livelihood. However, by paying premiums, farmers can transfer the risk of crop losses to insurance companies and receive compensation for their losses. We witnessed the struggles and successes of farmers and saw how they utilized different inputs for various crops and how they dealt with crop loss through insurance claims.

BEHIND THE HARVEST: CHALLENGES IN FARMING ECONOMICS

In states like Telangana, Karnataka, and Andhra Pradesh, where we did our research, crop loss was one of the biggest concerns for farmers. In the past few years, chilli and paddy crops have suffered major losses due to bad weather, resulting in a significant drop in production. We all knew about chilli farmers, but cotton farmers have also faced similar problems, with pests and unpredictable weather ruining their harvests. As researchers, we wanted to understand these challenges faced by farmers and find ways to help them overcome the challenges. So, we all sat together, surrounded by stacks of survey forms and notes, and started analysing the agricultural yield loss experienced by farmers in the past two years.

In dryland areas, all farmers experienced paddy crop loss in both years, with the average area under loss increasing from 3 to 6 acres (Table 13.1). This added to their financial strain. The story was similar for cotton farmers. They faced similar challenges, with high production costs leading to losses. However, chilli farmers were hit hardest, especially in 2021–22, due to increased pest infestation. These farmers were at the mercy of nature, and it seemed like nature was against them (Table 13.2). On the contrary, irrigated areas generally experienced lower crop losses. But, even there, production costs were a concern. It was clear that no matter the region or farming method, farmers were facing challenges in their yield and loss.

We were curious about these findings and decided to understand why there was such a significant increase in yield loss in the dryland areas and what impact it had on the farmers' income. We tried to remember that the fields we visited were not as abundant as they appeared. The crops were infested with pests and diseases. We spoke to one of the farmers sitting with us, and he mentioned the use of high concentrations of pesticides to combat pests and diseases, which had increased production costs. This, coupled with the decrease in total production, had a severe impact on their income. He also added that due to the increase in production costs and decrease

TABLE 13.1
Farmers Reported Yield Loss and Profitability in 2020–2021

Crop Names	Irrigation Status	Mean/ Number	Area (acres)	Total Production (q)	Yield (q/acre)	Total Cost of Cultivation (Rs.)	Price Received (Rs/q)	% Yield Loss Perception	Total Income (Rs.)	Total Profit/ Loss (Rs.)
Cotton	Rainfed	Mean	4.6	26	5.7	1,66,233	5,321	61	1,44,909	−26,507
		N	373				357	265	357	372
	Irrigated	Mean	3.3	23	7.1	1,11,816	5,566	53	1,32,093	20,720
		N	247				241	139	242	247
	Total	Mean	4.1	25	6.1	1,44,554	5,420	58	1,39,731	−7,662
		N	620				598	404	599	619
Paddy	Rainfed	Mean	3.9	92	23.6	98,354	1,906	34	1,61,167	56,045
		N	96				95	40	95	96
	Irrigated	Mean	5.4	123	22.7	1,38,940	1,753	37	1,97,213	62,172
		N	498				498	174	497	495
	Total	Mean	5.2	118	22.6	1,32,380	1,777	37	1,91,428	61,177
		N	594				593	214	592	591
Chillies	Rainfed	Mean	3.1	35	11.3	2,81,856	11,163	61	4,27,479	1,24,512
		N	202				189	102	189	202
	Irrigated	Mean	3.1	33	10.7	2,48,807	12,420	52	4,30,784	1,74,055
		N	394				385	166	384	392
	Total	Mean	3.1	34	10.9	2,59,908	12,001	55	4,29,259	1,56,909
		N	597				575	269	574	595
Redgram	Rainfed	Mean	4.1	12	3.0	65,370	5,201	66	73,569	568
		N	212				196	137	197	210
	Irrigated	Mean	3.9	18	4.5	60,230	5,417	58	1,08,612	42,042
		N	74				67	48	68	71
	Total	Mean	4.0	14	3.4	63,904	5,255	64	82,269	10,955
		N	287				264	186	266	282

(Continued)

TABLE 13.1 *(Continued)*
Farmers Reported Yield Loss and Profitability in 2020–2021

Crop Names	Irrigation Status	Mean/ Number	Area (acres)	Total Production (q)	Yield (q/acre)	Total Cost of Cultivation (Rs.)	Price Received (Rs/q)	% Yield Loss Perception	Total Income (Rs.)	Total Profit/ Loss (Rs.)
Maize	Rainfed	Mean	3.7	70	18.8	95,263	1,685	50	1,16,719	23,802
		N	57				57	20	57	56
	Irrigated	Mean	3.3	73	22.1	72,719	1,930	30	1,69,968	1,01,137
		N	57				55	20	55	57
	Total	Mean	3.5	71	20.4	83,991	1,805	40	1,42,868	62,812
		N	114				112	40	112	113
Chickpea	Rainfed	Mean	5.2	42	8.0	98,174	4,780	54	1,30,236	32,975
		N	46				46	21	46	46
	Irrigated	Mean	9.1	57	6.3	1,52,929	4,814	40	8,96,506	1,02,514
		N	14				14	6	14	14
	Total	Mean	6.1	45	7.4	1,10,950	4,788	51	3,09,032	49,201
		N	60				60	27	60	60
Tobacco	Rainfed	Mean	4.3	27	6.3	2,03,372	8,907	52	2,49,012	52,233
		N	43				43	12	43	43
	Irrigated	Mean	3.2	15	4.7	1,11,000	9,050	45	1,40,250	29,250
		N	10				10	4	10	10
	Total	Mean	4.0	25	6.2	1,85,943	8,934	50	2,28,491	47,896
		N	53				53	16	53	53
Sorghum	Rainfed	Mean	3.2	29	9.0	59,778	2,594	44	63,356	17,047
		N	18				18	5	18	17
	Irrigated	Mean	5.5	55	10.1	82,806	2,323	40	1,22,200	40,719
		N	31				31	11	31	31
	Total	Mean	4.7	46	9.7	74,347	2,422	41	1,00,584	32,335
		N	49				49	16	49	48

(Continued)

TABLE 13.1 *(Continued)*
Farmers Reported Yield Loss and Profitability in 2020–2021

Crop Names	Irrigation Status	Mean/ Number	Area (acres)	Total Production (q)	Yield (q/acre)	Total Cost of Cultivation (Rs.)	Price Received (Rs/q)	% Yield Loss Perception	Total Income (Rs.)	Total Profit/ Loss (Rs.)
Groundnut	Rainfed	Mean	2.8	24	8.4	49,167	6,583	10	1,12,583	64,500
		N	6				6	1	6	6
	Irrigated	Mean	2.7	22	8.2	56,571	3,876	41	95,630	37,053
		N	35				33	14	33	34
	Total	Mean	2.7	22	8.3	55,488	4,292	39	98,238	41,170
		N	41				39	15	39	40
Blackgram	Rainfed	Mean	4.2	18	4.3	1,04,444	5,394	58	1,42,856	−2,456
		N	18				16	6	16	18
	Irrigated	Mean	4.0	21	5.2	98,750	5,388	60	1,26,375	1,350
		N	10				8	4	8	10
	Total	Mean	4.1	19	4.6	1,02,411	5,392	59	1,37,363	−1,096
		N	28				24	10	24	28
Others	Rainfed	Mean	3.1	36	11.7	1,16,618	5,167	66	1,46,845	18,309
		N	70				60	30	60	67
	Irrigated	Mean	3.4	211	62.0	1,34,125	5,894	45	1,89,891	35,044
		N	60				55	18	55	57
	Total	Mean	3.2	118	36.9	1,24,824	5,514	58	1,67,432	26,002
		N	130				115	48	115	124
Total	Rainfed	Mean	4.0	34	8.5	1,51,108	5,901	59	1,83,204	24,489
		N	1,141				1,083	639	1,084	1,133
	Irrigated	Mean	4.1	71	17.3	1,54,039	5,863	47	2,46,592	84,015
		N	1,430				1,397	604	1,397	1,418
	Total	Mean	4.1	55	13.3	1,52,708	5,881	53	2,18,795	57,518
		N	2,573				2,482	1,245	2,483	2,553

Note: N indicates the number of farmers who reported crop loss.

TABLE 13.2

Farmers Reported Yield Loss and Profitability in 2021–2022

Crop Names	Irrigation Status	Mean/ Number	Area (acre)	Total Production (q)	Yield (q/ acre)	Cost of Cultivation (Rs.)	Price Received (Rs/q)	% Yield Loss Perception	Total Income (Rs.)	Total Profit/Loss (Rs.)
Cotton	Rainfed	Mean	4.1	21	5.0	1,45,540	6,466	50	1,41,064	−8,533
		N	377	377		377	363	377	377	376
	Irrigated	Mean	3.3	18	5.5	1,09,920	7,026	39	1,28,417	19,603
		N	226	224		226	218	226	226	224
	Total	Mean	3.8	20	5.2	1,32,190	6,676	46	1,36,324	1,971
		N	603	601		603	581	603	603	600
Paddy	Rainfed	Mean	4.0	79	19.8	1,03,645	1,703	25	1,33,977	33,266
		N	100	99		100	97	100	100	99
	Irrigated	Mean	5.7	118	20.8	1,46,793	1,755	19	1,91,015	44,299
		N	463	462		463	460	463	463	462
	Total	Mean	5.4	112	20.6	1,39,129	1,746	20	1,80,884	42,352
		N	563	561		563	557	563	563	561
Chillies	Rainfed	Mean	3.3	15	4.5	2,96,476	12,026	67	2,81,725	−1,12,936
		N	225	222		225	177	225	225	221
	Irrigated	Mean	3.2	10	3.2	2,59,742	13,953	66	1,42,711	−1,07,527
		N	421	420		421	290	421	421	412
	Total	Mean	3.2	12	3.7	2,72,502	13,211	66	1,91,019	−1,09,448
		N	647	643		647	468	647	647	634
Redgram	Rainfed	Mean	4.1	10	2.4	67,594	6,135	53	60,664	−3,367
		N	213	210		213	156	213	213	208
	Irrigated	Mean	4.1	14	3.3	61,459	6,324	53	80,036	23,066
		N	74	72		74	63	74	74	71

(Continued)

TABLE 13.2 *(Continued)*
Farmers Reported Yield Loss and Profitability in 2021–2022

Crop Names	Irrigation Status	Mean/ Number	Area (acre)	Total Production (q)	Yield (q/ acre)	Cost of Cultivation (Rs.)	Price Received (Rs/q)	% Yield Loss Perception	Total Income (Rs.)	Total Profit/Loss (Rs.)
	Total	Mean	4.1	11	2.7	66,012	6,189	53	65,659	3,360
		N	287	282		287	219	287	287	279
Maize	Rainfed	Mean	4.0	72	18.0	1,01,228	1,754	21	1,25,103	22,660
		N	50	50		50	49	50	50	49
	Irrigated	Mean	3.2	62	19.5	70,932	1,973	20	1,11,589	46,158
		N	59	59		59	57	59	59	58
	Total	Mean	3.6	69	19.2	86,785	1,871	21	1,21,626	37,292
		N	110	110		110	107	110	110	108
Chickpea	Rainfed	Mean	5.3	37	7.0	1,02,143	4,798	28	1,25,160	26,497
		N	49	49		49	48	49	49	47
	Irrigated	Mean	9.7	61	6.3	1,81,200	4,879	25	2,85,547	1,04,347
		N	15	15		15	14	15	15	15
	Total	Mean	6.3	43	6.8	1,20,672	4,816	27	1,62,751	45,331
		N	64	64		64	62	64	64	62
Tobacco	Rainfed	Mean	4.6	31	6.7	2,22,273	9,297	9	2,90,512	68,239
		N	33	33		33	33	33	33	33
	Irrigated	Mean	2.6	13	5.0	91,667	8,125	28	1,09,444	17,778
		N	9	9		9	8	9	9	9
	Total	Mean	4.2	27	6.5	1,94,286	9,068	13	2,51,712	57,426
		N	42	42		42	41	42	42	42
Sorghum	Rainfed	Mean	4.0	41	10.2	63,458	3,657	24	94,854	31,396
		N	24	24		24	23	24	24	24
	Irrigated	Mean	6.1	56	9.2	79,808	2,587	30	1,19,963	43,342

(Continued)

TABLE 13.2 *(Continued)*

Farmers Reported Yield Loss and Profitability in 2021–2022

Crop Names	Irrigation Status	Mean/ Number	Area (acre)	Total Production (q)	Yield (q/ acre)	Cost of Cultivation (Rs.)	Price Received (Rs/q)	% Yield Loss Perception	Total Income (Rs.)	Total Profit/Loss (Rs.)
		N	26	26		26	23	26	26	25
	Total	Mean	5.1	49	9.6	71,960	3,122	27	1,07,911	37,491
		N	50	50		50	46	50	50	49
Groundnut	Rainfed	Mean	2.5	23	9.4	42,143	6,871	3	1,24,429	1,82,286
		N	7	7		7	7	7	7	7
	Irrigated	Mean	3.3	27	8.3	72,214	4,189	13	1,22,193	53,907
		N	28	28		28	27	28	28	28
	Total	Mean	3.2	27	8.3	66,200	4,741	11	1,22,640	79,583
		N	35	35		35	34	35	35	35
Blackgram	Rainfed	Mean	6.0	16	2.6	1,51,630	5,532	57	93,322	−58,304
		N	23	23		23	19	23	23	23
	Irrigated	Mean	3.3	11	3.2	82,292	10,445	45	61,075	−21,217
		N	12	12		12	11	12	12	12
	Total	Mean	5.1	14	2.7	1,27,857	7,333	53	82,266	−45,589
		N	35	35		35	30	35	35	35
Total	Rainfed	Mean	4.0	28	7.0	1,51,339	6,534	46	1,52,801	−16,110
		N	1,176	1,165		1,175	1,038	1,176	1,176	1,160
	Irrigated	Mean	4.2	59	14.1	1,62,792	6,154	39	1,52,782	−5,559
		N	1,396	1,386		1,394	1,220	1,396	1,396	1,374
	Total	Mean	4.1	45	11.0	1,57,645	6,328	43	1,52,928	−10,337
		N	2,574	2,553		2,571	2,260	2,574	2,574	2,536

Note: N indicates the number of farmers who reported crop loss.

in total production, the farmers were not getting a fair price for their crops. The traders were also aware of the yield loss and took advantage of the situation by offering lower prices to the farmers. So it was clear that farmers were in a vicious cycle that needed to be broken.

INPUT USE PATTERN

When we revisited the farmers' homes to learn about their input usage habits, we were totally shocked. Although it was true that using proper and efficient inputs helps crops grow healthy and increase profit, many farmers (Narayanamoorthy et al., 2017) were using excessive pesticides and fertilizers on their chilli crops, thinking it would lead to bigger harvests and more profit (Table 13.3). But, sadly, the reality was quite different. Despite their efforts, their costs went up, and profits did not follow it. And to make matters worse, when chilli yields dropped, farmers suffered massive losses, sometimes crossing a lakh rupee per acre. There was a huge contrast to the more stable returns from paddy farming. After witnessing this reality, we realized how important sustainable farming practice is.

JOURNEY OF HARVESTED CROP

After learning about the reality of input usage, we wanted to shift our focus to something interesting. So we decided to explore how farmers utilized their crop output throughout the year. We focused on the four main crops grown in this area – paddy, maize, sorghum, and chickpea.

Nowadays more people produce crops for different purposes. Some crops are used as fodder, making biofuels (Ray et al., 2022), while paddy, being the most produced crop, was mostly used for household consumption as it served as a staple food for farmers and their families (Pingali, 2006). Sorghum and groundnut were also consumed at home but in smaller quantities. However, what surprised us was the limited amount of produce stored for seed purposes for the next season (Table 13.4).

TABLE 13.3
Input Usage by Farmers for Different Crops

Crop	State	Area (acre)		Seed (kg)		Urea (bags)		DAP (bags)		No. of sprays		Yield(q)		Price (Rs/q)	
		2021	2021	2021	2022	2021	2022	2021	2022	2021	2022	2021	2022	2021	2022
Cotton	Telangana	4.3	4.2	1	1	2.5	2.5	1.6	1.5	7	7	5.9	4.6	5986	6682
	Karnataka	4	3.9	1	1	1.8	1.9	2.1	2	6	6	7.6	6.3	5026	6842
	AP	4.9	4	1	1	2	2	1.8	1.8	6	6	5.9	5.2	5438	5998
	All	4.4	4	1	1	2	2.1	1.9	1.8	6	6	6.5	5.5	5421	6433
Paddy	Telangana	4.1	4.2	27	27	2.9	2.9	1.7	1.8	4	4	22.8	17.5	1967	2031
	Karnataka	9.7	9.5	27	27	2.6	2.6	2.4	2.3	2	2	24.2	22.7	1412	1470
	AP	4.7	4.6	28	28	2.6	2.5	2.2	2.2	2	2	21.3	20.3	1564	1636
	All	6.3	6.2	27	27	2.8	2.8	2	2	3	3	23.1	19.8	1699	1768
Chillies	Telangana	2.3	2.4	0.5	0.6	3.8	3.9	2.5	2.6	20	28	13.9	4.8	12290	14120
	Karnataka	5.6	6	0.6	0.5	2.6	2.6	3.8	3.8	12	12	10.4	4.9	13695	13650
	AP	2.5	2.6	0.5	0.6	2.8	2.8	3.1	3.1	12	12	9.4	4.8	9908	11315
	All	3.1	3.4	0.5	0.5	3.2	3.3	3	3	16	19	11.8	4.8	11998	13084
Maize	Telangana	2.3	2.4	8	8	1.7	2	1.5	1.2	3	3	22.3	15.8	1758	1810
	Karnataka	3.3	3.3	6	6	2.3	2	2.3	2.5	3	3	19.7	19.5	2067	2250
	AP	5.2	6.5	8	8	2.1	2.1	1.7	1.6	3	3	19.6	21.8	1702	1758
	All	4.2	5	8	8	2	2.1	1.7	1.6	3	3	20.4	19.8	1767	1824

TABLE 13.4
Utilization Patterns of Different Crops in the Study Area

Crop Names	Production (qtls)	Household Consumption (qtls)	Stored for Seed (qtls)	Marketable Surplus (qtls)	Marketed Quantity (qtls)	Marketed Quantity as % of Production
Cotton	22.2	0	0	22.2	22.2	100
Paddy	118	6.5	0	111.5	111.5	94.5
Chillies	21.5	0.2	0	21.3	21.3	99.1
Red gram	12.2	0.4	0.1	11.7	11.7	95.9
Maize	79.3	0.2	0	79.1	79.1	99.7
Chickpea	31.3	0.3	0	31	31	99.0
Tobacco	25.9	0	0	25.9	25.9	100
Sorghum	48.4	2.5	0	45.9	45.9	94.8
Groundnut	23.1	0.5	0.4	22.2	22.2	96.1
Black gram	18.9	0.3	0.1	18.5	18.5	97.9

Groundnut, red gram, and black gram were the only crops that were stored for this purpose, but, even then, it was not a significant amount. Most farmers said that they sold their crops immediately after harvesting either at local markets or through agencies. This was mainly because they did not have proper storage facilities, and storing crops was expensive. The transportation journey was also long, tiring, and incredibly challenging, but they had no other choice.

Through this chapter, we felt a sense of fulfilment and gained a deeper understanding of the farming business. Behind every harvest, there is a story, and farming is an ever-changing industry that requires constant adaptation and resilience. From that perspective, we tried to shed light on the struggles of farmers and help find solutions to the pressing issues they face. Now it is up to us to understand and support the farming business that sustains us all.

REFERENCES

Deshpande RS, Arora S. (Eds.). (2010). *Agrarian Crisis and Farmer Suicides* (Vol. 12). SAGE Publications Ltd.

Huet EK, Adam M, Giller KE, Descheemaeker K. (2020). Diversity in Perception and Management of Farming Risks in Southern Mali. *Agricultural Systems*, 184, Article no. 102905. https://doi.org/10.1016/j.agsy.2020.102905.

Mishra S (2008). Risks, Farmers' Suicides and Agrarian Crisis in India: Is There a Way Out?. Indian Journal of Agricultural Economics, 63(1), pp. 38–54.

Narayanamoorthy A. (2006). State of India's Farmers. *Economic and Political Weekly*, 41, pp. 471–473.

Narayanamoorthy A. (2007). Turnaround in Financial Recovery in Maharashtra's Irrigation Sector. *Economic and Political Weekly*, 42, pp. 2679–2685.

Narayanamoorthy A. (2013). Profitability in Crops Cultivation in India: Some Evidence from Cost of Cultivation Survey Data. *Indian Journal of Agricultural Economics*, 68(1), pp. 104–121.

Narayanamoorthy A, Suresh R, Alli P. (2017). Dynamics of Input Use Efficiency in Selected Crops Cultivation in India: A Temporal and Spatial Analysis. *Indian Journal of Agricultural Economics*, 72(3), pp. 215–229.

NCF. (2006). *Serving Farmers and Saving Farming, Report V Excerpts*. The National Commission on Farmers, Ministry of Agriculture, Government of India, New Delhi.

Pingali P. (2006). Westernization of Asian Diets and the Transformation of Food Systems: Implications for Research and Policy. *Food Policy*, 32(3), pp. 281–298.

Ray DK, Sloat LL, Garcia AS, Davis KF, Ali T, Xie W. (2022). Crop Harvests for Direct Food Use Insufficient to Meet the UN's Food Security Goal. *Nature Food*, 3(5), pp. 367–374.

Reddy AA, Raju SS, Bose A. (2020). Farmers' Income, Indebtedness and Agrarian Distress in India. *The Microfinance Review,* 12(1), pp. 20–38.

Ridley M, Rao G, Schilbach F, Patel V. (2020). Poverty, Depression, and Anxiety: Causal Evidence and Mechanisms. *Science*, 370(6522), eaay0214.

Sainath P. (2010), Farm Suicides – A 12 Year Saga. The Hindu, January 25.

Savary S, Willocquet L, Pethybridge SJ, Esker P, McRoberts N, Nelson A. (2019). The Global Burden of Pathogens and Pests on Major Food Crops. Nature Ecology & Evolution, 3(3), pp. 430–439.

14 Bridging the Gaps
Government Welfare Schemes and Their Reach

What we understood from talking to farmers was that they were not really worried but also psychologically and financially stressed about not having enough protection when their crops failed. This is a big issue, especially in developing countries where farmers cannot afford modern farming methods (Shankarnarayan and Ramakrishna, 2020) or adopt climate-resilient practices. Though Indian government supported farmers' livelihoods by increasing access to loans and offering tax relief (Harris et al., 2020), many farmers even with crop insurance, faced big losses they could not bounce back from. So, we decided to figure out what help they could get. As we looked into it more, we found out about other government programmes that might be able to assist these farmers. We spoke with local government bodies and did our own secondary research to learn more. And what we found was eye-opening. Since the agricultural sector supported 58% of the Indian population (Manida and Nedumaran, 2020), the government realized how important farmers are for the country's development and started a mission to improve their lives. They had come up with various schemes to tackle the many challenges farmers face and ensure their socio-economic empowerment (Table 14.1). Welfare schemes aimed at alleviating distress among farmers in India have played a critical role, but their functioning has often been marked by challenges in implementation and accessibility. The Pradhan Mantri Fasal Bima Yojana (PMFBY), launched in 2016, is one of the key initiatives aimed at mitigating risks from crop failure by offering insurance to farmers. However, while PMFBY has been instrumental in extending coverage to millions, issues like delayed claim settlements and lack of awareness among farmers have hampered its effectiveness (Kaur et al., 2021). Similarly, the Pradhan Mantri Kisan Samman Nidhi (PM-KISAN), which provides direct income support to small and marginal farmers, has been lauded for its immediate relief benefits, yet concerns over beneficiary targeting and exclusion errors persist. Despite these schemes, structural issues like inefficient bureaucracy, corruption, and the exclusion of tenant farmers continue to limit their impact. Moreover, welfare programmes such as the Mahatma Gandhi National Rural Employment Guarantee Act have provided supplemental income, helping reduce economic pressure on distressed farming households, but uneven implementation and delays in wage payments have reduced their effectiveness (Dreze and Khera, 2017). Overall, while welfare schemes have had some success in providing a safety net, more robust mechanisms are needed to address the deep-rooted financial vulnerabilities faced by Indian farmers.

DOI: 10.1201/9781003582878-14

TABLE 14.1
List of Welfare and Development Schemes Considered in the Analysis

Scheme Name	Eligibility	Benefit/Amount
Ration card	1. The applicant must be a resident of India. 2. The applicant should belong to a poor or economically weaker section of the society. 3. Newly married couples are eligible to apply for a ration card.	1. Used in the distribution of rice, white, sugar, and other applicable materials. 2. Provides free medical assistance in Arogyasri Hospitals.
PM-KISAN	1. Only small and marginal farmers whose land size is ≤ 5 acres are eligible for PM-KISHAN. 2. Farmer families that hold cultivated land can apply for the benefits of this plan. 3. A beneficiary should be an Indian citizen.	Rs. 6,000
Government health card (Aarogyasri Health Card)	Below-poverty-line (BPL) families.	The scheme provides financial assistance for the cost of hospitalization, diagnostic tests, surgeries, and other medical procedures related to the treatment of the covered illnesses.
Rythu Bandhu	1. Telangana state farmers are eligible to get the benefit of Rythu Bandhu Scheme. 2. Farmers who win the farm plot will get benefits of this scheme. 3. Small and margin agricultural labourers are eligible for the Rythu Bandhu Scheme. 4. Pattadar registered under Forest Rights Record can apply for this scheme.	Rs. 10,000
YSR Rythu Bharosa	1. Farmers of Andhra Pradesh, who own cultivated land, are eligible under this scheme. 2. As per the government, even those cultivating on endowments/temples/inam lands are eligible.	Rs. 7,500
Widow/old age/disability pension	Old-age persons, both male and female, who are 60 years of age or above and are destitute. Widow: As per the Marriage Act 18 years and above, disabled persons having a minimum of 40% disability and no age limit.	Rs. 24,000 (TS), 30,000 (AP), 2,400 (Karnataka)
Rythu Bhima	1. The scheme is applicable only to the farmers. 2. To prove their residence, the farmer must possess a Domicile certificate of the state under his name. 3. The farmers must belong to the age group of 18–59 years. 4. The applicant must own some piece of agricultural land under their name.	Rs. 5,00,000

(Continued)

TABLE 14.1 (*Continued*)
List of Welfare and Development Schemes Considered in the Analysis

Scheme Name	Eligibility	Benefit/Amount
Assistance with children's education	All Telangana SC, ST, BC, OBC, minority, and Divyanga students.	Up to Rs. 35,000 only irrespective of their college fee (If the college fee is less, then the amount will be considered.)
Soil health cards	All farmers in India are eligible for the SHG scheme. They must contact the agriculture department in their region for more.	Rs. 190
Indira Avasa Yojana	BPL households living in the rural areas, belonging to SC/ST, minorities in the BPL category, and non-SC/ST BPL rural households and registered in BPL List 2002, ex-servicemen, and retired members of paramilitary forces fulfilling the other conditions.	Rs. 70,000
Mid-day meal	Mid-Day Meal Scheme by providing every child in every government and government-assisted primary school with a prepared mid-day meal.	1. Children from classes I–VIII will get a nutritious meal every day except the holidays. 2. This scheme ensures that children who do not get food at home will have at least one meal a day. 3. Children can get food for free.
Seed subsidy	All farmers	Depending on the crop from 30% to 50% of the seed cost
Ammavadi	1. Must be a legal resident of AP, must be from a BPL household with a white ration card, students study in between classes 1 and 12. If the child discontinues the study in the middle of the session of the academic year, then they will not be able to avail the benefits of the scheme.	Rs. 13,000
Kisan Credit Cards	1. Famers who are individuals or joint borrowers of the cultivated land and are engaged in farming or related activities. 2. Individual land owners as well as cultivators. 3. Tenant farmers, oral leases, and shared croppers of the cultivated lands. 4. Self-help groups or joint liability groups formed by sharecroppers or tenant farmers.	The Kisan Credit Card (KCC) Yojana provides short-term loans to farmers to meet their urgent expenses arising during cultivation and for maintaining their farming equipment. It allows farmers to avail of credits at low interest rates from banks and financial institutions.
Integrated Child Development Services (ICDS)/ Anganwadi	1. Children below six years, pregnant and lactating mothers (P&LM). 2. Women (15–45 years)	1. To improve the nutritional and health status of children in the age group of 0–6 years. 2. To reduce the incidence of mortality, morbidity, malnutrition, and school dropout. 3. Supplementary nutrition. 4. Pre-school non-formal education. 5. Immunization, health check-up, and referral services

(*Continued*)

TABLE 14.1 (*Continued*)
List of Welfare and Development Schemes Considered in the Analysis

Scheme Name	Eligibility	Benefit/Amount
YSR Cheyutha	1. Women from weaker sections of society such as SC/ST/OBC/minority community. 2. The applicant must be above the age of 45 years. An applicant must be below the age of 60 years	Rs. 18,750
Risk mitigating (farm pond construction, etc.)	Only those farmers or entrepreneurs having land ownership in the state shall be eligible for availing assistance under the scheme.	Rs. 75,000
KVK/SAUs/ ICAR	All farmers	Providing quality training to the farmers/farmwomen, also conducting frontline demonstrations on proven technologies
Kalyana Laxmi/ Gruha Laxmi	1. The applying girl should belong to any of the mentioned minority communities or should fall under the BPL. 2. The applicant should be a resident of Telangana state. 3. The unmarried girl must have completed 18 years of age at the time of marriage.	Rs. 1,00,116
NGOs		Some NGOs are working, but the impact is negligible.

Excited by this, we started looking into these schemes in detail. They covered a lot of things, from monetary help and crop insurance to making sure the soil is healthy and changing how things are sold. But what stood out the most were the benefits for families who were struggling. But it was not so easy, especially in a country like India where a lot of people are poor, hungry, and do not have jobs. The government had to figure out who was poor and how to help them properly.

It was really nice to see that the government was making sure nobody got left behind in their fight against poverty. There were subsidies for job opportunities, safety nets (Maiorano, 2014; Patnaik and Das, 2017), free education for children, pensions for old people and widows, and even land for those who don't have any (Debnath et al., 2018), cooking gas, free meals for school children, and many more.

But we wanted to make sure these programmes were truly helping the farmers who needed them the most. So, we kept looking into it, studying the schemes and how they were affecting the farmers' lives. But as we kept going, we found some gaps in the system (Acharya Balkrishna et al., 2022). While the central government was giving money for these programmes, it was up to the state governments to make them work in reality. And that was where things got tricky. Many state governments were not using the money properly, so the schemes were not reaching the people they were supposed to help. We knew that, if India wanted to get rid of poverty, the state governments had to step up and do their part. The government's idea of "putting people first" (Sachs, 2012) would only work if every state made sure these schemes were working well. We realized that bridging the gap between the central and state governments was crucial for the success of these welfare schemes. The money allocated for these programmes needed to be utilized properly and reach the people who needed it the most.

So, we went back to talking to the farmers to try and figure out how well these government programmes were really helping them out.

HOW DOES SUPPORT VARY AMONG DIFFERENT GROUPS?

In our study, we looked at how government schemes reached people during both a normal year in 2020–2021 and a bad one in 2021–2022 when farmers experienced a severe crop loss. We had an open discussion with the farmers to closely analyse the reach of these schemes in both years. Two schemes called Rythu Bandhu and Pradhan Mantri Kisan Samman Nidhi or PM-KISAN stood out at that time. These were the special programmes for farmers, giving them Rs. 5,000 per acre in each season and Rs. 6,000 in each year respectively. The interesting thing is, that more farmers benefited from Rythu Bandhu and PM-KISAN compared to any other agriculture-related schemes. This shows how much support farmers received during both good and tough times.

Religious Groups

The main schemes benefiting village farmers were Rythu Bandhu, PM-KISAN, and Rythu Bhima. But other schemes like ration cards, Arogya Sree, YSR Rythu Bharosa, and mid-day meals helped everyone, regardless of religion. Arogya Sree, especially, saw more Muslim and Hindu beneficiaries in 2021 (Table 14.2). Hindu and Muslim households had access to ICDS/Anganwadi, mid-day meals, and Kalyana Laxmi/Gruha Laxmi schemes but not Christian households. Overall, government schemes reached farmers and households of all religions in the village. Hindus benefited more from PM-KISAN and Rythu Bandhu, but Rythu Bhima expanded to include Christians and Muslims in 2021–2022 (Table 14.3). This showed efforts for equal support across religious groups. As the researcher, we were happy to see these schemes positively impacting farmers' lives.

Social Groups

It has been amazing to witness how these schemes have changed the farmer's lives. One big takeaway was how universal schemes like ration cards and PM-KISAN reached almost all households, no matter their social background. These schemes reached about 90% of the target households, regardless of their social background (Table 14.4; Figure 14.1). We noticed very little variation in the beneficiaries across social groups. This meant that socially backward groups were not excluded from these schemes, which was a huge relief. In fact, we found that most of the scheduled caste households were benefiting from schemes such as widow pensions, Indira Awas Yojana, and the mid-day meal scheme. Among all the social groups, we were happy to see that the scheduled tribe households were being

TABLE 14.2
Reach of Government Schemes by Religious Groups (Percent of Households) in Normal Year (2020–2021)

	Religious Groups			
Schemes	**Hindu**	**Muslim**	**Christian**	**Total**
Ration card	92.9	98.1	95.7	93.2
PM-KISAN	86.7	80.8	73.9	86.3
Government health card (AarogyaSree)	64.2	51.9	87.0	64.2
Rythu Bandhu	31.9	19.2	17.4	31.2
YSR Rythu Bharosa	29.3	17.3	60.9	29.4
LIC/other insurance	28.8	23.1	8.7	28.3
Widow/old age/disability pension	24.1	17.3	43.5	24.2
Assistance with children's education	15.7	15.4	8.7	15.6
Rythu Bhima	12.3	0.0	0.0	11.7
Indira Avasa Yojana	9.9	15.4	8.7	10.1
Soil health cards	10.0	5.8	8.7	9.9
Seed subsidy	8.8	15.4	0.0	8.9
Mid-day meal	8.2	13.5	4.3	8.3
Ammavadi	6.5	5.8	17.4	6.6
Kisan credit cards	5.8	7.7	0.0	5.8
ICDS/Anganwadi	3.9	5.8	0.0	3.9
YSR Cheyutha	2.2	0.0	0.0	2.0
Risk mitigating (farm pond construction, etc.)	1.4	0.0	0.0	1.3
Kalyana Laxmi/Gruha Laxmi	1.0	0.0	0.0	1.0
KVK/SAUs/ICAR	1.0	0.0	0.0	0.9
NGOs	0.2	0.0	0.0	0.2

TABLE 14.3
Reach of Government Schemes (Percent of Households) by Religion of Households in Bad Year (2021–2022)

	Religion Groups			
Schemes	**Hindu**	**Muslim**	**Christian**	**Total**
Ration card	93.0	98.1	91.3	93.2
PM-KISAN	86.6	80.8	73.9	86.2
Government health (Aarogya Sree)	66.6	53.8	87.0	66.5
Rythu Bandhu	31.5	19.2	17.4	30.8
YSR Rythu Bharosa	29.4	17.3	65.2	29.6
LIC/other insurance	29.1	21.2	8.7	28.5
Widow/old age/disability pension	24.1	17.3	43.5	24.2
Rythu Bhima	17.0	7.7	13.0	16.5
Assistance with children's education	15.9	15.4	8.7	15.8
Soil health cards	10.0	7.7	8.7	9.9

(Continued)

TABLE 14.3 (*Continued*)

Reach of Government Schemes (Percent of Households) by Religion of Households in Bad Year (2021–2022)

	Religious Groups			
Schemes	**Hindu**	**Muslim**	**Christian**	**Total**
Indira Avasa Yojana	9.5	15.4	8.7	9.7
Mid-day meal	9.0	13.5	4.3	9.1
Seed subsidy	9.0	15.4	0.0	9.1
Ammavadi	6.5	5.8	17.4	6.6
Kisan credit cards	5.9	7.7	0.0	5.8
ICDS/Anganwadi	4.2	5.8	0.0	4.2
YSR Cheyutha	2.2	0.0	0.0	2.0
Risk mitigating (farm pond construction, etc.)	1.6	0.0	0.0	1.5
KVK/SAUs/ICAR	1.3	0.0	0.0	1.3
Kalyana Laxmi/Gruha Laxmi	1.3	3.8	0.0	1.3
NGOs	0.2	0.0	0.0	0.2

TABLE 14.4

Government Scheme Beneficiaries (Percent of Households) by Social Group in Normal Year (2020–2021)

	Caste Groups				
Schemes	**SC**	**ST**	**OBC**	**General**	**Total**
Ration card	96.2	93.8	95.0	90.5	93.2
PM-KISAN	77.1	88.4	86.3	87.8	86.3
Government health card (AarogyaSree)	62.6	54.8	66.2	64.9	64.2
Rythu Bandhu	22.1	43.8	44.0	17.7	31.2
YSR Rythu Bharosa	25.2	8.2	18.3	46.3	29.4
LIC/other insurance	21.4	17.8	31.3	29.6	28.3
Widow/old age/disability pension	35.1	24.0	22.6	23.2	24.2
Assistance with children's education	16.8	14.4	16.5	14.6	15.6
Rythu Bhima	9.2	17.1	17.1	5.7	11.7
Indira Avasa Yojana	15.3	9.6	10.1	9.0	10.1
Soil health cards	4.6	15.1	11.6	8.1	9.9
Seed subsidy	6.9	8.9	10.9	7.4	8.9
Mid-day meal	11.5	11.6	9.4	5.7	8.3
Ammavadi	9.9	0.0	5.0	9.1	6.6
Kisan credit cards	3.1	8.9	8.0	3.4	5.8
ICDS/Anganwadi	3.8	6.8	5.5	1.5	3.9
YSR Cheyutha	3.1	0.0	2.3	2.1	2.0
Risk mitigating (farm pond construction, etc.)	0.8	1.4	2.7	0.2	1.3
Kalyana Laxmi/Gruha Laxmi	0.8	2.1	1.1	0.7	1.0
KVK/SAUs/ICAR	0.0	1.4	0.7	1.2	0.9
NGOs	0.0	0.0	0.5	0.0	0.2

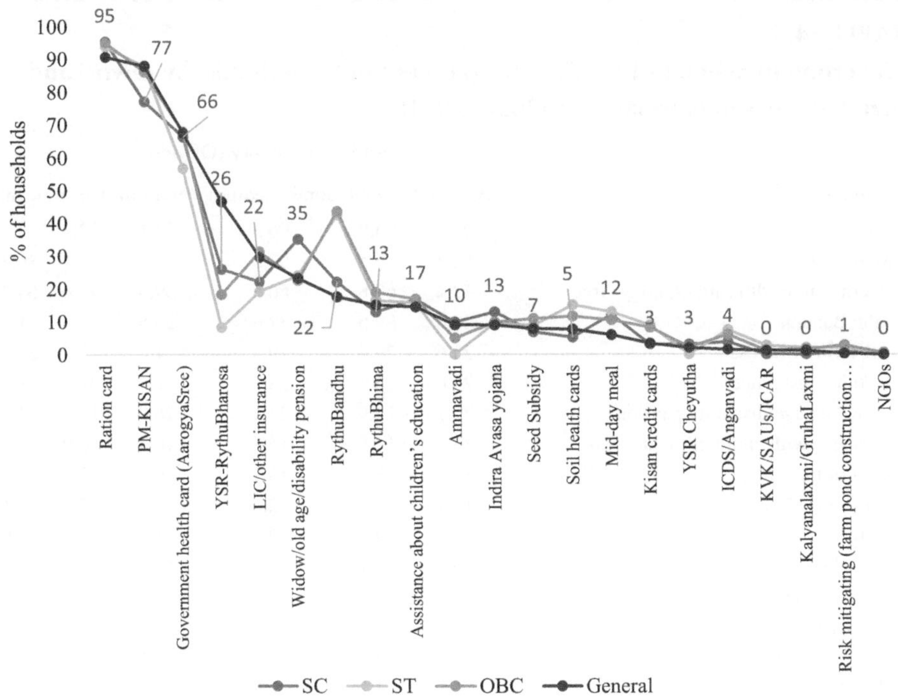

FIGURE 14.1 Percentage of beneficiaries from different schemes by social group in a bad year (2021–2022).

reached by schemes like Kisan Credit Cards and ICDS/Anganwadi. This means the government was trying hard to help the most vulnerable people be a part of these schemes.

But perhaps the best fact was that land-based schemes like PM-KISAN, YSR Rythu Bharosa, and Rythu Bandhu were reaching all social groups equally. So, we realized that these government schemes were not just giving financial support but also creating a sense of equality and unity among different social groups.

Landholding Groups (Own Land)

After talking with farmers for hours, we discovered that PM-KISAN, which helps small and marginal farmers, reached more of these farmers. This means the scheme was working to support the most vulnerable farmers. Other schemes like Aarogya Sree, pensions, children's educational help, ammavadi, ICDS/Anganwadi, and YSR Cheyutha were also helping more marginal landholders (Table 14.5; Figure 14.2). The farmers said that some schemes, like information sharing, farm pond construction, YSR Cheyutha, and ICDS/Anganwadi, were not reaching many of their households. This suggests the need for better ways to make these benefits known.

TABLE 14.5
Government Schemes Beneficiaries (Percent of Households) by Own-Land Size Category in Normal Year (2020–2021)

	Land Size Category (Owned)					
Schemes	**Marginal**	**Small**	**Semi-Medium**	**Medium**	**Large**	**Total**
Ration card	94.0	94.9	96.5	93.1	72.3	93.2
PM-KISAN	62.8	85.5	90.6	92.4	90.1	86.3
Government health card (AarogyaSree)	72.4	69.8	66.5	59.6	41.6	64.2
Rythu Bandhu	28.6	39.6	37.6	25.5	8.9	31.2
YSR Rythu Bharosa	25.6	29.4	29.5	31.4	27.7	29.4
LIC/other insurance	25.6	26.7	28.7	27.5	39.6	28.3
Widow/old age/disability pension	31.7	23.5	21.2	26.0	16.8	24.2
Assistance with children's education	21.6	14.5	14.7	16.4	6.9	15.6
Rythu Bhima	12.6	12.5	15.1	9.1	3.0	11.7
Indira Avasa Yojana	10.1	15.7	12.0	6.4	2.0	10.1
Soil health cards	5.0	9.0	10.9	11.8	8.9	9.9
Seed subsidy	6.0	5.1	8.1	10.8	19.8	8.9
Mid-day meal	11.1	13.7	7.0	5.6	5.9	8.3
Ammavadi	9.0	5.9	5.7	8.1	2.0	6.6
Kisan credit cards	2.5	5.9	5.0	7.1	9.9	5.8
ICDS/Anganwadi	6.0	4.7	2.8	3.4	4.0	3.9
YSR Cheyutha	4.0	1.6	1.8	2.0	1.0	2.0
Risk mitigating (farm pond construction, etc.)	1.0	1.2	1.8	1.2	1.0	1.3
Kalyanalaxmi/Gruha Laxmi	1.5	1.6	0.9	0.5	1.0	1.0
KVK/SAUs/ICAR	0.5	0.8	0.7	1.2	2.0	0.9
NGOs	0.5	0.0	0.2	0.2	0.0	0.2

Surprisingly, fewer large landowners were benefiting from schemes like ration cards and Aarogya Sree. This means these schemes were not reaching everyone, especially those with more land. But it was good to see that even land-based schemes like PM-KISAN and YSR Rythu Bharosa were helping the poorest farmers. This shows that the government is trying to support all farmers, not just certain ones.

Landholding Groups (Total Cultivated Land)

"*My family's livelihood was entirely dependent on our small plot of land, but due to the crazy weather and limited resources, our harvest was never enough to sustain us throughout the year. Then suddenly, the PM-KISAN scheme. I still remember I couldn't believe it when I got my first payment. It was a significant amount for my family, and it helped us buy better seeds and tools for our farm,*" said one marginal farmer when we asked them to share their experience on government schemes.

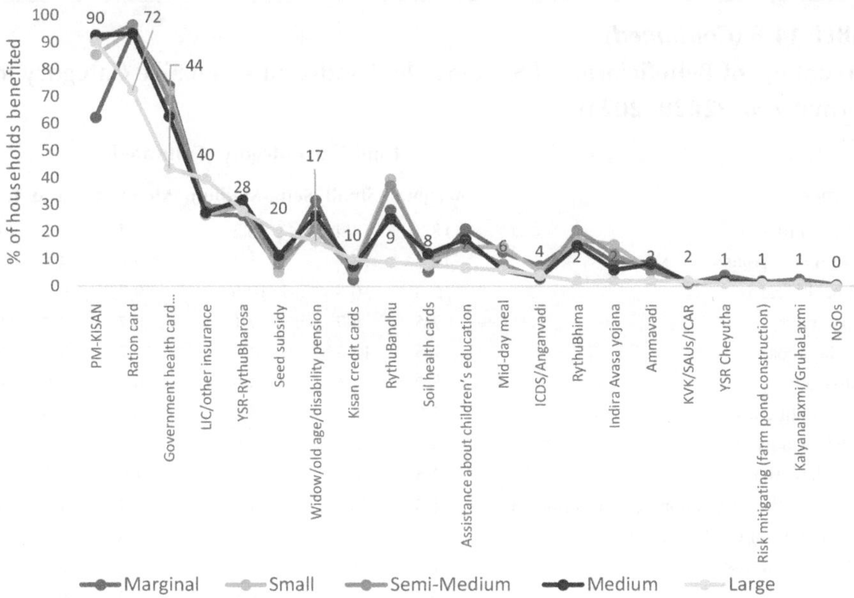

FIGURE 14.2 Percent beneficiaries under different government schemes by own-land size category in a bad year (2021–2022).

But that was not the only benefit they received from the government. The farmers, especially the smaller landholders, were also supported enough by schemes like ration cards and the Aarogya Sree scheme (Table 14.6; Figure 14.3). The village leaders also played important roles in getting the farmers enrolled in the land-based schemes. Farmers of this group were grateful for the government's

TABLE 14.6
Percentage of Beneficiaries of Schemes by Land-Cultivated Size Category in Normal Year (2020–2021)

	Own Land Size Category					
Schemes	**Marginal**	**Small**	**Semi-Medium**	**Medium**	**Large**	**Total**
Ration card	90.0	94.9	97.1	93.9	84.5	93.2
PM-KISAN	90.0	83.1	86.2	87.2	85.8	86.3
Government health card (Aarogya Sree)	72.5	66.2	66.3	63.4	57.3	64.2
Rythu Bandhu	48.8	40.0	38.6	26.1	15.9	31.2
YSR Rythu Bharosa	28.8	19.5	26.3	34.4	32.3	29.4
LIC/other insurance	18.8	23.6	25.1	29.1	39.7	28.3
Widow/old age/disability pension	32.5	23.1	25.6	24.3	19.4	24.2
Assistance with children's education	16.3	11.3	14.0	19.0	14.2	15.6

(Continued)

TABLE 14.6 (*Continued*)
Percentage of Beneficiaries of Schemes by Land-Cultivated Size Category in Normal Year (2020–2021)

	Land Size Category (Cultivated)					
Schemes	**Marginal**	**Small**	**Semi-Medium**	**Medium**	**Large**	**Total**
Rythu Bhima	18.8	14.4	15.2	9.7	5.2	11.7
Indira Avasa Yojana	13.8	13.3	11.3	9.9	4.3	10.1
Soil health cards	6.3	10.8	10.6	9.7	9.5	9.9
Seed subsidy	7.5	6.7	7.1	8.7	14.7	8.9
Mid-day meal	7.5	12.3	7.9	7.7	7.3	8.3
Ammavadi	5.0	2.6	4.9	10.1	6.0	6.6
Kisan credit cards	5.0	6.2	5.7	5.1	7.3	5.8
ICDS/Anganwadi	5.0	4.6	2.9	4.5	3.0	3.9
YSR Cheyutha	2.5	0.5	2.9	1.6	2.6	2.0
Risk mitigating (farm pond construction, etc.)	1.3	0.5	1.7	1.4	1.3	1.3
Kalyana Laxmi/Gruha Laxmi	2.5	1.5	1.0	0.8	0.4	1.0
KVK/SAUs/ICAR	1.3	1.0	0.5	1.2	0.9	0.9
NGOs	0.0	0.0	0.5	0.2	0.0	0.2

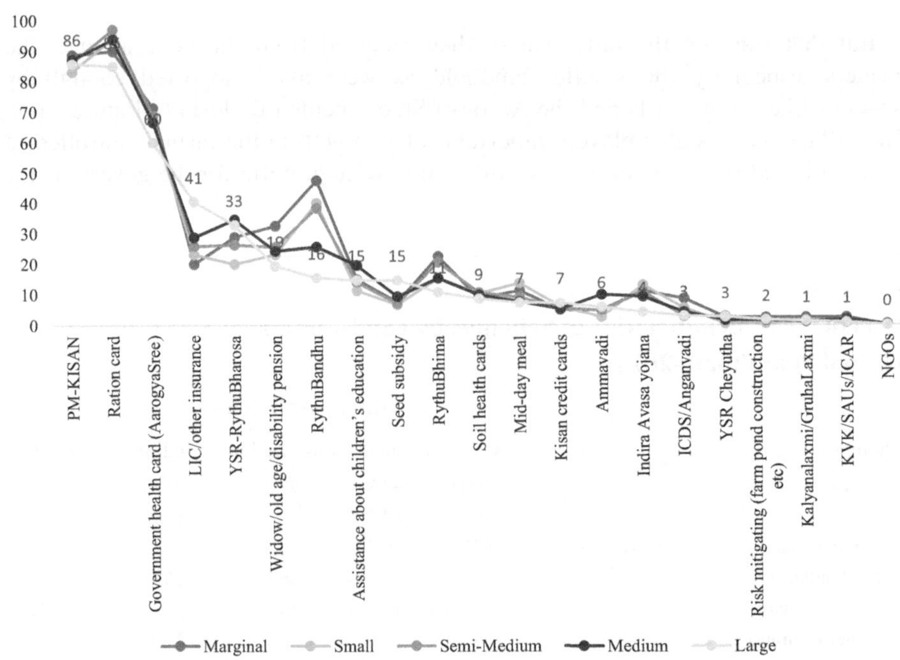

FIGURE 14.3 Percentage of beneficiaries under different government schemes by land-cultivating class (2021–2022).

efforts as those truly made a difference in their lives. It truly made a difference in their lives. We were delighted to see how these government's schemes reached even the neediest.

Poverty Groups

But our excitement quickly turned into disappointment as we spoke to farmers from impoverished backgrounds. Despite the government's efforts to tackle poverty and improve living conditions, the non-poor farmers were getting more benefits from these schemes compared to the poor farmers (Table 14.7; Figure 14.4) – the exact opposite of what should have happened. We inspected further and discovered several hurdles that prevented the poorest from availing the benefits of these schemes. Many of the poorest had no idea how to apply for these schemes or the benefits they could avail. Even some of them did not hear about the schemes. Further, the complex application process, coupled with corrupted middlemen, created barriers to accessing benefits. Additionally, poor infrastructure like limited transportation and internet

TABLE 14.7
Reach of Government Schemes by Poverty Status (Percentage of Households Benefited)

	Poverty Status					
	2020-21			2021-22		
Schemes	Non-poor	Poor	Total	Non-poor	Poor	Total
Ration card	92.9	93.6	93.2	93.8	92.5	93.2
PM-KISAN	86.6	85.6	86.3	86.6	85.9	86.3
Government health card (Aarogya Sree)	63.2	66.1	64.2	62.1	66.5	64.2
Rythu Bandhu	36.7	20.1	31.2	31.9	30.4	31.2
YSR Rythu Bharosa	25.1	37.9	29.4	26.6	32.5	29.4
LIC/other insurance	32.6	19.7	28.3	30.3	26.0	28.3
Widow/old age/disability pension	25.8	20.8	24.2	26.0	22.1	24.2
Assistance with children's education	16.5	13.8	15.6	15.6	15.5	15.6
Rythu Bhima	14.3	6.4	11.7	12.1	11.3	11.7
Indira Avasa Yojana	11.5	7.2	10.1	11.0	9.0	10.1
Soil health cards	10.9	7.8	9.9	9.1	10.7	9.9
Seed subsidy	8.0	10.6	8.9	9.1	8.6	8.9
Mid-day meal	7.7	9.5	8.3	8.3	8.3	8.3
Ammavadi	6.6	6.6	6.6	6.6	6.6	6.6
Kisan credit cards	6.6	4.0	5.8	6.8	4.7	5.8
ICDS/Anganwadi	4.3	3.0	3.9	3.7	4.1	3.9
YSR Cheyutha	2.0	2.1	2.0	2.8	1.2	2.0
Risk mitigating (farm pond construction, etc.)	2.0	0.0	1.3	0.8	2.0	1.3
Kalyanalaxmi/Gruha Laxmi	1.1	0.8	1.0	1.3	0.6	1.0
KVK/SAUs/ICAR	1.2	0.4	0.9	0.5	1.4	0.9
NGOs	0.3	0.0	0.2	0.1	0.3	0.2

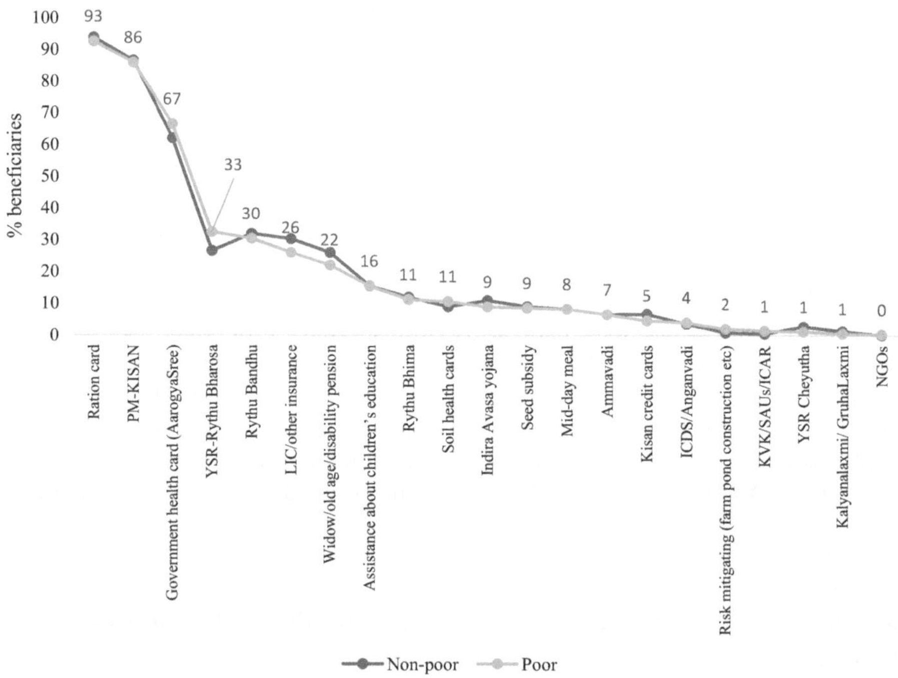

FIGURE 14.4 Percentage of beneficiaries under different government schemes by poor and non-poor in a bad year (2021–2022).

connectivity made it difficult for the poorest to reach the offices and apply for these schemes.

Tenancy Groups

We felt really disheartened to see the same tough situation among the farmers in this group. In all the schemes, the number of tenant farmers getting benefits was way lower than that of owner farmers (Table 14.8; Figure 14.5). Even though programmes like ICDS/Anganwadi and mid-day meals were more beneficial to tenant farmers as they targeted the poorest segments of society, tenant farmers were often deprived of support from land-focused schemes that they needed the most. This was because most of these schemes were linked to land ownership, which the tenant farmers did not have. This lack of land ownership has always been a big problem for them in getting help from government schemes. They did not have any documents to prove their rights as tenants, which was usually needed for many schemes. Expressing deep disappointment some farmers said – "*Without land, we can't get help from these schemes, and without the schemes, we can't make our lives better.*"

Besides the other obstacles, the biggest challenge for the tenant farmers was the lack of targeted schemes that cater to their specific needs, because most of the schemes were designed for landowners, leaving out the marginalized community.

TABLE 14.8
Government Schemes Beneficiaries (Percentage of Households) by Land Tenancy Status

	2020–2021			2021–2022		
Schemes	Owner Farmer	Tenant Farmer	Total	Owner Farmer	Tenant Farmer	Total
Ration card	93.2	93.1	93.2	93.2	93.1	93.2
PM-KISAN	89.9	18.1	86.3	89.8	18.1	86.2
Government health card (AarogyaSree)	63.9	68.1	64.2	66.2	72.2	66.5
Rythu Bandhu	32.3	9.7	31.2	32.0	9.7	30.8
YSR Rythu Bharosa	30.6	5.6	29.4	30.8	6.9	29.6
LIC/other insurance	27.8	37.5	28.3	28.0	38.9	28.5
Widow/old age/disability pension	24.1	25.0	24.2	24.1	25.0	24.2
Assistance with children's education	15.1	23.6	15.6	17.1	5.6	16.5
Rythu Bhima	12.2	2.8	11.7	15.4	23.6	15.8
Indira Avasa Yojana	10.3	5.6	10.1	10.1	5.6	9.9
Soil health cards	10.1	5.6	9.9	9.9	5.6	9.7
Seed subsidy	9.0	6.9	8.9	8.8	13.9	9.1
Mid-day meal	8.0	13.9	8.3	9.2	6.9	9.1
Ammavadi	6.7	5.6	6.6	6.7	5.6	6.6
Kisan credit cards	5.9	2.8	5.8	6.0	2.8	5.8
ICDS/Anganwadi	3.6	8.3	3.9	3.9	8.3	4.2
YSR Cheyutha	2.0	2.8	2.0	2.0	2.8	2.0
Risk mitigating (farm pond construction, etc.)	1.3	1.4	1.3	1.6	1.4	1.5
Kalyanalaxmi/Gruha Laxmi	0.9	2.8	1.0	1.3	0.0	1.3
KVK/SAUs/ICAR	1.0	0.0	0.9	1.3	2.8	1.3
NGOs	0.1	1.4	0.2			

Irrigation Groups

Surprisingly, both dryland and irrigated farmers had similar access to government benefits. However, after digging deeper, we found some hidden challenges for dryland farmers. We found that the government gave hope to dryland farmers. Many dryland farmers benefited from schemes like Rythu Bharosa and Rythu Bandhu, which provided financial help and support. Though Rythu Bandhu mostly covered the irrigated farmers, both schemes together made a big difference in dryland farmers' lives (Table 14.9; Figure 14.6).

One such farmer we met during our research had a small piece of land that he inherited from his father, but it was in a drought-prone area. With no irrigation facilities, he relied on unpredictable monsoons, which often led to crop failures and financial strain. He had to take loans and borrow money from relatives to sustain his family. But with Rythu Bharosa's help, he invested in irrigation and better seeds, resulting in a successful harvest and improved livelihood. But we also witnessed that despite government support many dryland farmers still faced struggles due to factors like inadequate rainfall and poor soil quality.

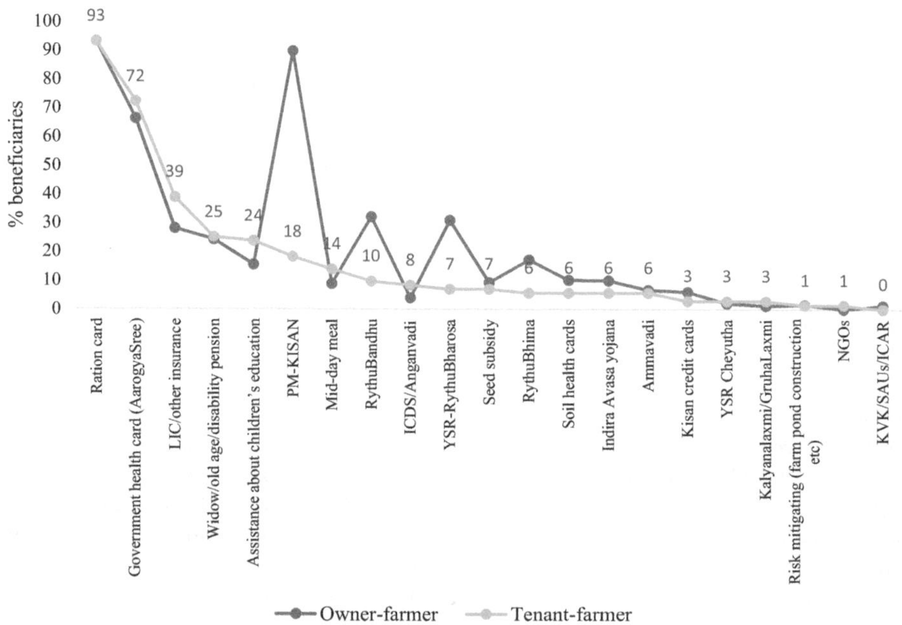

FIGURE 14.5 Percentage of beneficiary households under different government schemes by land tenancy status in a bad year (2021–2022).

TABLE 14.9
Percentage of Beneficiaries (Households) under Government Schemes by Dryland and Irrigated Farmers

Schemes	2020–2021 Dry Land	2020–2021 Irrigated Land	2020–2021 Total	2021–2022 Dry Land	2021–2022 Irrigated Land	2021–2022 Total
Ration card	93.9	92.7	93.2	94.1	92.6	93.2
PM-KISAN	80.1	90.1	86.3	80.3	89.8	86.2
Government health card (Aarogya Sree)	71.0	59.9	64.2	73.9	61.9	66.5
Rythu Bandhu	9.4	44.7	31.2	9.2	44.3	30.8
YSR Rythu Bharosa	49.6	16.8	29.4	50.0	16.9	29.6
LIC/other insurance	24.4	30.7	28.3	24.8	30.8	28.5
Widow/old age/disability pension	28.5	21.5	24.2	28.5	21.5	24.2
Assistance with children's education	17.6	14.3	15.6	13.1	18.7	16.5
Rythu Bhima	3.3	16.9	11.7	17.3	14.8	15.8
Indira Avasa Yojana	8.8	10.8	10.1	8.3	10.8	9.9
Soil health cards	8.3	10.8	9.9	8.8	10.3	9.7
Seed subsidy	9.4	8.6	8.9	7.5	10.0	9.1
Mid-day meal	7.5	8.8	8.3	9.6	8.8	9.1

(*Continued*)

TABLE 14.9 (*Continued*)
Percentage of Beneficiaries (Households) under Government Schemes by Dryland and Irrigated Farmers

	2020–2021			2021–2022		
Schemes	Dry Land	Irrigated Land	Total	Dry Land	Irrigated Land	Total
Ammavadi	12.9	2.7	6.6	12.7	2.9	6.6
Kisan credit cards	4.6	6.5	5.8	4.6	6.6	5.8
ICDS/Anganwadi	3.1	4.3	3.9	3.1	4.8	4.2
YSR Cheyutha	3.5	1.1	2.0	3.5	1.1	2.0
Risk mitigating (farm pond construction, etc.)	0.4	1.9	1.3	0.4	2.3	1.5
Kalyana Laxmi/Gruha Laxmi	1.1	0.9	1.0	0.0	2.1	1.3
KVK/SAUs/ICAR	0.0	1.5	0.9	1.1	1.5	1.3
NGOs	0.2	0.2	0.2	0.2	0.2	0.2

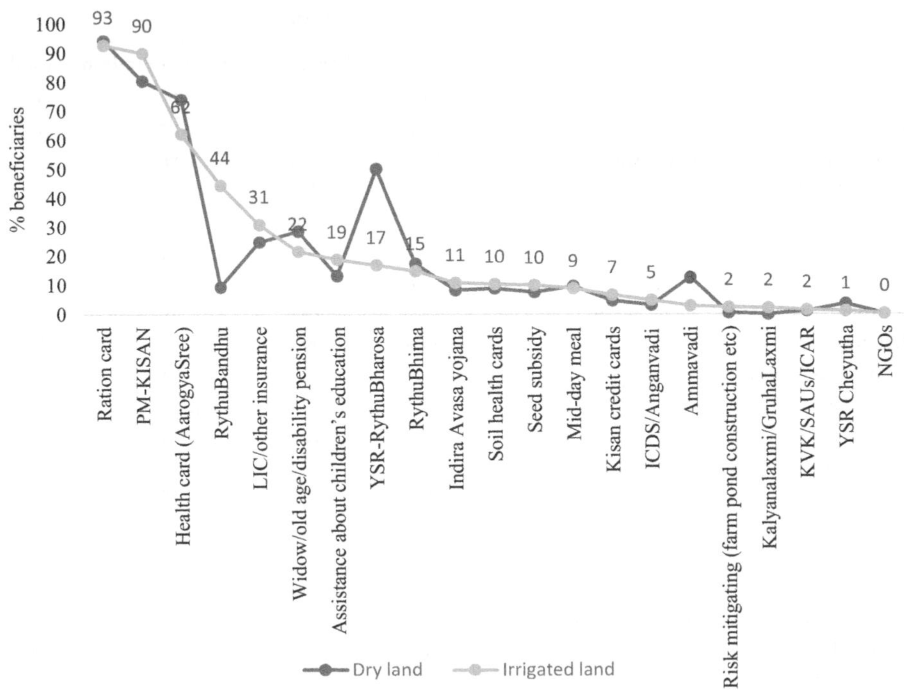

FIGURE 14.6 Percentage of beneficiaries of government schemes among dryland and irrigated farmers in a bad year (2021–2022).

COMPARISONS AMONG STATES

As we compared the states, we wondered how many people were left out or included wrongly in government schemes. Inclusion errors happened when data was wrong or rules were not followed properly, while exclusion errors occurred when people were missed out due to lack of outreach or bureaucratic issues. When we discussed our findings with our team, we saw that some schemes were for everyone and used by all states, but others were only for certain places. Even among the universal schemes, the number of people benefiting varied from state to state. For instance, the PM-KISAN scheme, meant for all farmers, had more beneficiaries in Telangana and Andhra Pradesh (Table 14.10; Figure 14.7). We found out that some schemes targeted specific groups, while others were for families below the poverty line.

So what we found out was that, even though a lot of money goes into farming in India, many farmers still struggle. The help from government schemes did not reach

TABLE 14.10
Percent Beneficiaries under Different Government Schemes by State in Normal Year (2020–2021)

	State Name			
Schemes	**Telangana**	**Karnataka**	**Andhra Pradesh**	**Total**
Ration card	91.1	98.3	90.0	93.2
PM-KISAN	88.5	87.5	82.9	86.3
Government health card (AarogyaSree)	75.9	30.2	86.9	64.2
Rythu Bandhu	95.9	NA	0.4	31.2
YSR RythuBharosa	NA	NA	86.9	29.4
LIC/other insurance	33.0	24.8	27.3	28.3
Widow/old age/disability pension	19.8	18.3	34.2	24.2
Assistance with children's education	15.2	10.6	20.8	15.6
Rythu Bhima	35.9	NA	0.2	11.7
Indira Avasa yojana	15.4	6.7	8.3	10.1
Soil health cards	14.1	11.3	4.4	9.9
Seed subsidy	2.0	22.3	2.1	8.9
Mid-day meal	9.3	10.4	5.2	8.3
Ammavadi	NA	NA	19.6	6.6
Kisan credit cards	7.8	8.1	1.5	5.8
ICDS/Anganwadi	7.0	3.3	1.5	3.9
YSR Cheyutha	NA	NA	6.0	2.0
Risk mitigating (farm pond construction, etc.)	3.3	0.8	0.0	1.3
Kalyanalaxmi/Gruha Laxmi	1.3	1.7	0.0	1.0
KVK/SAUs/ICAR	2.0	0.6	0.2	0.9
NGOs	0.7	0.0	0.0	0.2

NA, not applicable.
The schemes are not implemented in these states; they are specific to a particular state.

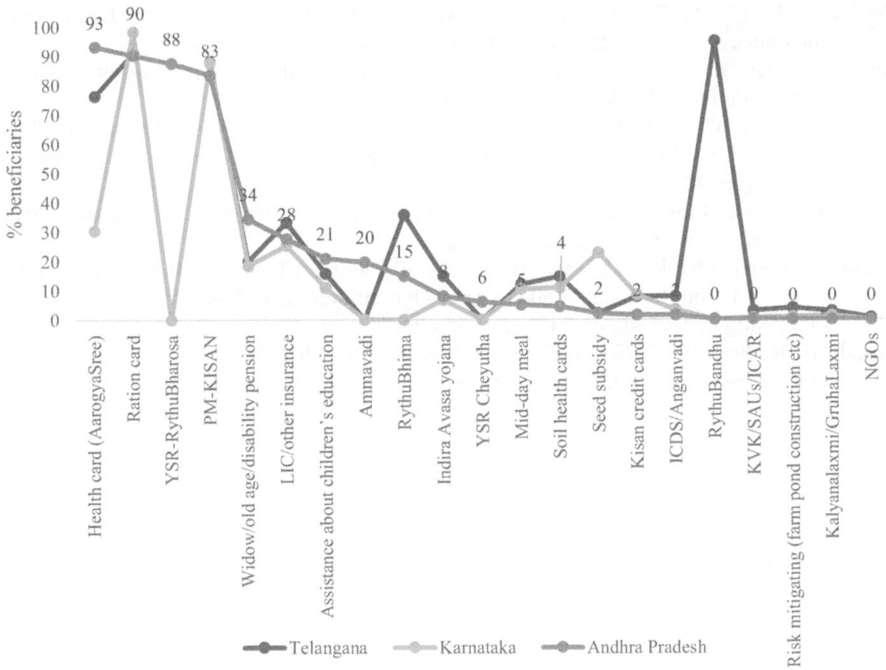

FIGURE 14.7 State-wise percentage of beneficiaries under different schemes (2021–2022).

the farmers who needed it most (Prabha et al., 2016). What we figured out was that there was a need to create awareness among farmers through training and awareness of extension services to get maximum benefits and help in achieving the schemes for the farmers (Singh and Rathore, 2023).

REFERENCES

Acharya Balkrishna AK, Arya V, Sharma G, Srivastava D, Sharma A, Khan N. (2022). Situational Analysis of Government Initiatives for the Welfare of Farmers in India: Impact and Futuristic Insights. *Indian Journal of Ecology*, 49(3), pp. 1023–1035.

Debnath D, Babu S, Ghosh P, Helmar, M. (2018). The Impact of India's Food Security Policy on Domestic and International Rice Market. *Journal of Policy Modeling*, 40(2), pp. 265–283.

DrezeJ, Khera R. (2017). Recent Social Security Initiatives in India. *World Development,* 98, 555–572.

Manida M, Nedumaran DG (2020). Agriculture in India: Information about Indian Agriculture and Its importance. *Agaeum Journal*, 8(3), pp. 729–736.

Harris J, Depenbusch L, Pal AA, Nair RM, Ramasamy S. (2020). Food System Disruption: Initial Livelihood and Dietary Effects of COVID-19 on Vegetable Producers in India. *Food Security*, 12(4), pp. 841–851.

Kaur S, Raj H, Singh H, Chattu V K. (2021). Crop Insurance Policies in India: An Empirical Analysis of the Pradhan Mantri Fasal Bima Yojana. *Risks*, 9 (11), 191.

Maiorano D. (2014). The Politics of the Mahatma Gandhi National Rural Employment Guarantee Act in Andhra Pradesh. *World Development*, 58, pp. 95–105.

Patnaik U, Das PK. (2017). Do Development Interventions Confer Adaptive Capacity? Insights from Rural India. *World Development*, 97, pp. 298–312.

Prabha RK, Rai BNJP, Singh SR. (2016). Role of Government Schemes in Indian Agriculture and Rural Development. In Goyal SK, Rai JP, Kumar S (Eds.), *Indian Agriculture and Farmer* (pp. 92–102). Poddar Publication.

Sachs JD. (2012). From Millennium Development Goals to Sustainable Development Goals. *The Lancet*, 379(9832), pp. 2206–2211.

Shankarnarayan VK, Ramakrishna H. (2020). Paradigm Change in Indian Agricultural Practices Using Big Data: Challenges and Opportunities from Field to Plate. *Information Processing in Agriculture*, 7 (3), pp. 355–368.

Singh M, Rathore R. (2023). Agricultural Schemes in India: A Review. *Asian Journal of Agricultural Extension, Economics and Sociology*, 41(9), pp. 730–736.

15 Effects of Farmer Loan Waivers

We gained a lot of firsthand experience of the extensive struggles of the farmers. Although listening to these struggles was very painful, it was necessary to know these struggles for our research. We learned that, when natural calamities struck or crop prices came down, farmers were left with no choice but to take out loans to survive. We heard about the loan waiver schemes in India, so we had a quick glance at our secondary research. We found that the scheme was first implemented nearly three decades ago in early 1990 (Pathak and Chattopadhyay, 2021), promising to pay off crop loans up to Rs. one lakh for small farmers (Tripathi, 2020) and supposed to provide relief to distressed farmers who took a loan from formal credit institutions but were unable to repay their loans due to various reasons (Narayanan and Mehrotra, 2019). Since then, both central and state governments have implemented similar schemes. The schemes might seem like a lifeline for distressed farmers, but, in reality, they came with a heavy price. These loan waivers put a strain on the economy, and the funds allocated for them could have been used for other development projects. But the impact goes beyond just the economy. The political parties also tended to use it as a tactic to come to power (Jha et al., 2019). Loan waivers also created a moral hazard and actually made things worse. When farmers saw they could get their loans forgiven, some of them started skipping out on paying back what they owed. It became this never-ending loop of debt, and they relied more and more on the government for help. The worst part was it messed up the whole system of borrowing money, and farmers had a harder time getting loans after that. The inequity in the selection criteria for loan waivers also created resentment and division among farmers. In a word, loan waivers offered only temporary relief without addressing the underlying issues like low productivity and poor infrastructure and did not solve the bigger problems. As a result, there was not much benefit to marginal farmers who actually needed the scheme (Chawla and Rout, 2021). So, in the last stage of our fieldwork, we decided to delve deeper into the facts and tried to understand the negative consequences of these loan waivers.

HOW DO LOAN WAIVERS IMPACT VARIOUS GROUPS OF FARMERS?

Religious Groups

When we spoke to the farmers from different religious backgrounds, we observed that all the beneficiaries of loan waivers predominantly belonged to the Hindu religion (Table 15.1). Even the disparity continued in the subsequent year. We were shocked and decided to go further.

DOI: 10.1201/9781003582878-15

TABLE 15.1
Beneficiaries of Loan Waive Off by Religion (Rs.)

	Religion			
	Hindu		Total	
Source of Credit	**Mean**	**N (%)**	**Mean**	**N (%)**
Commercial banks (2020–2021)	40,929	1	40,929	1
Commercial banks (2021–2022)	48,500	0	48,500	0
Regional rural bank (2020–2021)	18,250	0	18,250	0
Co-operative societies/banks (2020–2021)	55,333	0	55,333	0
Co-operative societies/banks (2021–2022)	50,000	0	50,000	0
Traders (2020–2021)	30,000	0	30,000	0
Total loan waiver (formal, 2021)	42,522	2	42,522	2
Total loan waiver (informal, 2021)	30,000	0	30,000	0
Total loan waiver (formal, 2022)	49,000	0	49,000	0

N, % of households. "0" indicates less than 1% of households.

We asked them about their experiences with the loan waiver scheme and how it had impacted their lives. To our surprise, the answers were similar. Most farmers belonging to minority religious groups had not received any loan waivers. Not only that, even among Hindu farmers, merely 1% benefitted from the scheme. We understood that it was not a coincidence. It was a clear indication of the biases and discrimination that existed in the implementation of the scheme where minority religious groups were being left behind.

Social Groups

Conversely, despite being a marginalized community, scheduled caste (SC) and scheduled tribe (ST) categories received loan waivers from various banks. However, the number of beneficiaries was nominal, and that too dropped down in the following year (Table 15.2). This sudden drop in loan waivers came as a shock to all of them who were getting benefits. A Muslim expressed his disappointment, saying, "*We lost our hopes, and we wondered why this was happening. We were promised support and relief, but instead, we were left to face the harsh reality.*" It was clear that farmers were frustrated and betrayed by the system.

Landholding Groups (Own Land)

The loan waiver amount for small and marginal farmers was a big hope as it was more than enough to pay off all their debts and start afresh. But, in reality, they met with disappointment; not even 3% of them had received the waiver (Table 15.3). For other farmers, it was even worse. We had no words to answer when some marginal farmers howled to us, "*Will we ever receive the support and benefits we deserve?*" But, as researchers, we clearly understood that these schemes were not meant to

TABLE 15.2
Beneficiaries of Loan Waive Off by Social Category (Rs.)

	Caste							
	SC		ST		OBC		General	
Source of Credit	Mean	N	Mean	N	Mean	N	Mean	N
Commercial banks (2020–2021)	29,500	3	50,000	1	48,333	1	32,500	0
Commercial banks (2021–2022)	20,000	1		0	20,000	0	77,000	0
Regional rural bank (2020–2021)		0	25,000	1	11,500	0		0
Co-operative societies/banks (2020–2021)		0		0	45,500	1	75,000	0
Co-operative societies/banks (2021–2022)		0		0	50,000	0	50,000	0
Traders (2020–2021)		0		0	30,000	0		0
Total loan waiver (formal, 2021)	29,500	3	37,500	3	45,000	2	53,750	1
Total loan waiver (informal, 2021)		0		0	30,000	0		0
Total loan waiver (formal, 2022)	20,000	1		0	35,000	0	68,000	1

N, % of households. Zero indicates less than 1% of households.

TABLE 15.3
Beneficiaries of Loan Waive Off by Own Land Size Category (Rs.)

	Marginal		Small		Semi-Medium		Medium	
Source of Credit	Mean	N	Mean	N	Mean	N	Mean	N
Commercial banks (2020–2021)	39,000	3	50,000	2	24,000	0	30,000	0
Commercial banks (2021–2022)	1,00,000	1	20,000	1		0	54,000	0
Regional rural bank (2020–2021)		0	25,000	1	11,500	0		0
Co-operative societies/banks (2020–2021)	1,00,000	1	50,000	0	45,500	1		0
Co-operative societies/banks (2021–2022)		0	50,000	0	50,000	0		0
Traders (2020–2021)	20,000	1		0	40,000	0		0
Total loan waiver (formal, 2021)	49,167	3	50,000	3	31,625	2	30,000	0
Total loan waiver (informal, 2021)	20,000	1		0	40,000	0		0
Total loan waiver (formal, 2022)	1,00,000	1	30,000	1	50,000	0	54,000	0

uplift the farmers genuinely; instead, they seemed to aim at appeasing the farmers for some political gain.

Landholding Groups (Total Cultivated Land)

Surprisingly, one fact that caught our attention was that the majority of loan waivers were given to farmers with smaller cultivated land sizes, and commercial banks took the lead on this (Table 15.4). To know a little more about this, we further interviewed

TABLE 15.4
Beneficiaries of Loan Waive Off by Cultivated Land Category

	Total Cultivated Land Wise 2022									
	Marginal		Small		Semi-Medium		Medium		Large	
Source of Credit	**Mean**	**N**	**Mean**	**N**	**Mean**	**N**	**Mean**	**N**	**Mean**	**N**
Commercial banks (2020–2021)	40,000	5	42,500	2	42,667	1	32,500	0	50,000	0
Commercial banks (2021–2022)		0	60,000	1	20,000	0	54,000	0		0
Regional rural bank (2020–2021)		0	25,000	1	11,500	1		0		0
Co-operative societies/banks (2020–2021)		0		0	50,000	1	75,000	0	41,000	1
Co-operative societies/banks (2021–2022)		0		0	50,000	0		0	50,000	0
Traders (2020–2021)		0		0		0	30,000	0		0
Total loan waiver (formal, 2021)	40,000	5	36,667	3	35,857	2	53,750	1	66,000	1
Total loan waiver (informal, 2021)		0		0		0	30,000	0		0
Total loan waiver (formal, 2022)		0	60,000	1	35,000	1	54,000	0	50,000	0

some of the farmers and came to know that smaller farmers got loan waivers from regional rural banks (RRBs), whereas co-operative societies/banks were found to give loan waivers mostly to the larger farmers. We understood that the loan waiver scheme had a greater impact on farmers with smaller landholdings compared to those with larger ones.

Poverty Groups

In both the poverty groups, below poverty line (BPL) and above poverty line, only a tiny 1% of farmers received loan waivers from commercial banks (Table 15.5). This meant that most poor farmers, who should have benefitted most, were left out. What was even more surprising was that non-poor farmers received much higher loan waivers than the poor. This was a big change from the previous year when most waivers went to the poor. We were puzzled by this and decided to understand the reason behind this trend. The shocking truth we uncovered was corruption. It broke our hearts to see these poor farmers unfairly left out of the loan waiver scheme.

Tenancy Groups

Once again, we observed a similar pattern: only a mere 1% of farmers who were owner-farmers received the benefit of loan waivers from commercial banks. As we delved further, we found out that the situation was not much different for pure tenants, with not even 1% of them receiving loan waivers from co-operative societies/banks (Table 15.6). Farmers shared that co-operative societies provided waivers only to owner-farmers in 2021–2022, leaving out tenants once more. We really wondered how this scheme, which was meant to help all farmers, was only benefiting a select few.

TABLE 15.5
Beneficiaries of Loan Waive Off by Poverty Status of Households (Rs.)

	Poverty Status, 2021						Poverty Status, 2022					
	Non-poor		Poor		Total		Non-poor		Poor		Total	
Source of Credit	**Mean**	**N**	**Mean**	**N**	**Mean**	**N**	**Mean**	**N**	**Mean**	**N**	**Mean**	**N**
Commercial banks (2020–2021)	42,538	1	20,000	0	40,929	1	46,000	1	38,111	1	40,929	1
Commercial banks (2021–2022)	37,000	0	60,000	0	48,500	0	77,000	0	20,000	0	48,500	0
Regional rural bank (2020–2021)	17,667	0	20,000	0	18,250	0	20,000	0	16,500	0	18,250	0
Co-operative societies/ banks (2020–2021)	55,333	1		0	55,333	0	46,400	1	1,00,000	0	55,333	0
Co-operative societies/ banks (2021–2022)	50,000	0		0	50,000	0	50,000	0		0	50,000	0
Traders (2020–2021)	20,000	0	40,000	0	30,000	0	30,000	0		0	30,000	0
Total loan waiver (formal, 2021)	44,667	2	20,000	0	42,522	2	45,636	1	39,667	2	42,522	2
Total loan waiver (informal, 2021)	20,000	0	40,000	0	30,000	0	30,000	0		0	30,000	0
Total loan waiver (formal, 2022)	43,500	0	60,000	0	49,000	0	63,500	1	20,000	0	49,000	0

N, % of total households; zero indicates less than 1%.

TABLE 15.6
Beneficiaries of Loan Waive Off by Tenancy Status (Rs.)

	Owner-Farmer		Tenant-Farmer	
Source of Credit	**Mean**	**N**	**Mean**	**N**
Commercial banks (2020–2021)	40,929	1		0
Commercial banks (2021–2022)	31,333	0	1,00,000	1
Regional Rural Bank (2020–2021)	18,250	0		0
Co-operative societies/banks (2020–2021)	46,400	0	1,00,000	1
Co-operative societies/banks (2021–2022)	50,000	0		0
Traders (2020–2021)	40,000	0	20,000	1
Total loan waiver (formal, 2021)	39,909	2	1,00,000	1
Total loan waiver (informal, 2021)	40,000	0	20,000	1
Total loan waiver (formal, 2022)	38,800	0	1,00,000	1

Irrigation Groups

However, we were surprised to see that co-operative societies/banks showed a greater inclination towards granting loan forgiveness to dryland cultivators in 2020–2021 (Table 15.7). But the most interesting finding was regarding RRBs. They displayed a stronger preference for extending loan waivers exclusively to irrigated farmers. When we spoke with both dryland and irrigated farmers, we noticed that, while the loan waiver scheme had helped some dryland farmers, it had also been misused by others. Hardly 1.5% of farmers who received the benefit of loan waiver from commercial banks had irrigated land in 2020–2021. This was a tiny fraction compared to the total number of farmers being granted waivers. However, things took a little change in 2021–2022, and there was a substantial drop in loan waivers granted by commercial banks for irrigated farmers. However, we were surprised to find that co-operative societies and banks showed a stronger inclination towards granting loan forgiveness to dryland cultivators in 2020–2021. The most intriguing discovery was about RRBs. They seemed to favour exclusively irrigated farmers.

COMPARISON AMONG STATES

We were quite surprised to find out that only 3% of farmers in Telangana had received loan waivers from commercial banks, and even fewer from RRBs and co-operative societies (Table 15.8). The situation was even worse in the following year. What shocked us even more was that the states of Karnataka and Andhra Pradesh hadn't received any loan waivers at all.

It was disheartening to see the despair in their eyes as they shared their stories with us. It was obvious that these loan waivers were not reaching the most vulnerable and marginalized farmers. But we were glad we figured out the main problem: many poor farmers could not get the waivers because they did not have anything to offer as collateral (Reddy et al., 2022). Most of these farmers were already struggling, and, without collateral, they could not qualify for the waivers.

TABLE 15.7
Beneficiaries of Loan Waive Off by Irrigation Status (Rs.)

Source of Credit	Dry Land		Irrigated Land	
	Mean	N	Mean	N
Commercial banks (2020–2021)	18,000	0	42,692	2
Commercial banks (2021–2022)	1,00,000	0	31,333	0
Regional Rural Bank (2020–2021)		0	18,250	1
Co-operative societies/banks (2020–2021)	66,000	0	50,000	1
Co-operative societies/banks (2021–2022)		0	50,000	0
Traders (2020–2021)	30,000	0		0
Total loan waiver (formal, 2021)	50,000	1	41,400	2
Total loan waiver (informal, 2021)	30,000	0		0
Total loan waiver (formal, 2022)	1,00,000	0	38,800	1

TABLE 15.8
Beneficiaries of Loan Waived Off by State (Rs.)

	State Name							
	Telangana		Karnataka		Andhra Pradesh		Total	
Source of Credit	**Mean**	**N**	**Mean**	**N**	**Mean**	**N**	**Mean**	**N**
Commercial banks (2020–2021)	40,929	3.0		0.0		0.0	40,929	1.0
Commercial banks (2021–2022)	48,500	0.9		0.0		0.0	48,500	0.3
Regional rural bank (2020–2021)	18,250	0.9		0.0		0.0	18,250	0.3
Co-operative societies/banks (2020–2021)	55,333	1.3		0.0		0.0	55,333	0.4
Co-operative societies/banks (2021–2022)	50,000	0.4		0.0		0.0	50,000	0.1
Traders (2020–2021)		0.0	40,000	0.2	20,000	0.2	30,000	0.1
Total loan waiver (formal, 2021)	42,522	5.0		0.0		0.0	42,522	1.6
Total loan waiver (informal, 2021)		0.0	40,000	0.2	20,000	0.2	30,000	0.1
Total loan waiver (formal, 2022)	49,000	1.3		0.0		0.0	49,000	0.4

REFERENCES

Chawla R, Rout CK. (2021). Farm Loan Waiver: Need for Framers Upliftment in View Agricultural Scholars. *AgriCos e-Newsletter*, 2(2), pp. 8–12.

Jha S, Mohapatra AK, Lodha SS. (2019). Political Economy of Farm Loan Waivers in India. *FIIB Business Review*, 8(2), pp. 88–93.

Narayanan S, Mehrotra N. (2019). Loan Waivers and Bank Credit: Reflections on the Evidence and the Way Forward. *Vikalpa*, 44(4), pp. 198–210.

Pathak A, Chattopadhyay AK. (2021). Debates on Agricultural Loan Waiver Schemes in India: Myths and Realities. *Contemporary South Asia*, 29(4), pp. 560–570.

Reddy AA, Sarkar A, Onishi Y. (2022). Assessing the Outreach of Targeted Development Programmes—A Case Study from a South Indian Village. *Land*, 11(7), p. 1030.

Tripathi GC. (2020). *Impact Evaluation of Farm Debt Waiver Scheme on Farmers' Livelihood in Uttar Pradesh*. Agro-Economic Research Centre, University of Allahabad. https://desagri.gov.in/wp-content/uploads/2024/03/Microsoft-Word-Final-Report-Debt-Waiver.pdf

16 Farmers Distress Index

We spent months in fieldwork to witness the struggles and hardships that farmers faced, from unpredictable weather to fluctuating crop prices and how their lives were full of uncertainty and distress. We collected data on various issues such as crop prices, access to resources, and indebtedness and considered the psychological impact of distress on farmers and their families. It was a daunting task, but, as researchers, we were determined to find a way to measure and understand the level of distress faced by farmers in India. The **Farmers Distress Index (FDI)** was developed as part of the research project "**Agrarian Distress and PMFBY: An Analysis of Rainfed Agriculture**" at **ICAR-CRIDA**, with funding from **NABARD**. This research initiative piloted the index, and a version of the FDI has been further developed into an **Android application**. The earlier version of FDI is published in Land journal (see the link https://www.mdpi.com/2073-445X/10/11/1236). Later, a simplified version was developed as web application. The application can be accessed at https://agripulse.info/.

The **Farmers Distress Index (FDI)** is a **user-friendly tool** designed to **forewarn stakeholders**, including the government, about the **severity of farmer distress** based on **seven key pillars**. It enables **timely preventive action** by identifying areas in need of intervention. The FDI also proposes a **scalable framework** for implementation at **sub-district levels**, ensuring that government support reaches the most affected regions efficiently. This index is developed based on **seven key pillars**: (i) **Exposure to risk, (ii) Adaptive capacity, (iii) Sensitivity to exposure, (iv) Adaptation, (v) Trigger, (vi) Psychological vulnerability of households, (vii) Impact.** From each pillar, **three indicator variables** are selected for tracking, resulting in a total of **21 indicators**. This structured approach enables **swift intervention** by identifying areas with **high distress scores**, ensuring timely support where it is most needed. Table 16.1 provided the details of the variables/indicators (Reddy et al., 2021).

The **Farmer Distress Index (FDI)** is designed to develop a forewarning system for to take preventive measures to eliminate farmer distress, providing alerts **three months in advance** based on **21 leading indicators**. These indicators are drawn from **seven key pillars** of farmer distress, with three variables selected from each pillar for monitoring.

When signs of localized distress emerge, the local administration can administer a standardized **21-question survey** in the affected areas. This assessment enables authorities to diagnose the **severity of farmer distress on a 1–10 scale**, where **below 3 indicates low distress** and **above 7 signals high distress**. Additionally, the index identifies which of the **seven pillars—exposure, adaptive capacity, sensitivity, adaptation, trigger, psychological, or impact**—requires intervention.

By leveraging these insights, local administrations can implement **preventive measures** to mitigate farmer distress before it escalates into a widespread crisis.

DOI: 10.1201/9781003582878-16

WHAT MYSTERY DID THE FDI REVEAL?

We calculated the FDI based on various factors such as their religion, social group, land size, poverty status, irrigation status, and tenancy status. Though 2021–2022 was not an abnormal year in terms of weather, the severe pest attack on the chilli crop made the farmers unable to cope with the sudden changes in the environment and left them devastated. As a result, farmers suffered heavy losses due to their low adaptive capacity and adaptation.

As we got deeper into the concepts of adaptive capacity and adaptation, we realized that they were closely related but focused on different things. Adaptive capacity was the inherent ability of a system to adjust and cope with changes, while adaptation was the actual process of making changes in response to changing circumstances.

The findings prompted us to understand the root causes behind this. The three main indicators of adaptive capacity that we focused on were education level, land size, and leased-in land (Table 16.1). It was disheartening to discover that the education level of households among the sample farmers was quite low. This meant that they were unable to adapt quickly to pest attacks and often relied on pesticide dealers who recommended excessive doses, resulting in increased costs and low yields. Furthermore, the farmers in this region had small landholdings, which were not suitable for growing high-return and high-risk crops. This made them risk-averse and unable to take advantage of potentially more profitable opportunities. Additionally, many farmers had leased land to grow cash crops like chillies, which led to further losses due to the high rent and unpredictable market prices. To make matters worse,

TABLE 16.1
Explanation of Indicators Used in Farmers Distress Index (FDI)

Pillar	Indicator 1	Indicator 2	Indicator 3
Exposure	Loss due to pest/ diseases (%)	Floods/cyclones (%)	Droughts (%)
Adaptive capacity	Average education of the household (years)	Total owned land (acre)	Leased-in land (acre)
Sensitivity	Irrigated area (% of total area)	Indebtedness (Rs)	SC/SC community and number of children in the household
Adaptation	Non-crop income (as % of total household income)	Number of government schemes household benefited (number)	Household savings with self-help groups and cooperatives through membership (Rs.)
Trigger	Informal credit (Rs)	Pressure from repayment of loans (yes/no)	Lack of capital for investment in agriculture (yes/no)
Psychological	Feeling of social isolation (yes/no)	Unable to fulfil family obligations (yes/no)	Addicted to alcohol (yes/no)
Impact	Increased indebtedness (Yes/No)	More participation in public works (MGNREGA) (yes/no)	Reduced food consumption (yes/no)

Note: Variables highlighted in grey are inversely related to FDI.

the non-crop income of these farmers was very low, and they did not have access to sufficient government schemes that could alleviate their distress. Even when they did benefit from schemes like the Pradhan Mantri Fasal Bima Yojana, they often faced delays in receiving claims (Table 16.2).

TABLE 16.2
Farmers Distress Index (FDI) among Different Categories of Farmers

Religion	Exposure	Adaptive Capacity	Sensitivity	Adaptation	Trigger	Psychology	Impact	FDI
Hindu	0.17	0.56	0.26	0.63	0.35	0.25	0.25	0.35
Muslim	0.12	0.57	0.25	0.66	0.26	0.24	0.26	0.34
Christian	0.18	0.59	0.52	0.54	0.39	0.36	0.32	0.41
Social Group								
SC	0.17	0.56	0.45	0.60	0.37	0.26	0.31	0.39
ST	0.19	0.54	0.36	0.66	0.34	0.28	0.33	0.39
OBC	0.17	0.56	0.21	0.61	0.34	0.25	0.29	0.35
OC	0.16	0.57	0.25	0.64	0.35	0.25	0.20	0.35
Land Ownership Status								
Marginal	0.16	0.60	0.34	0.60	0.37	0.25	0.25	0.36
Small	0.17	0.58	0.28	0.60	0.35	0.28	0.28	0.36
Medium	0.17	0.57	0.26	0.63	0.33	0.26	0.29	0.36
Semi-medium	0.18	0.55	0.25	0.64	0.36	0.25	0.24	0.35
Large	0.15	0.46	0.24	0.70	0.33	0.20	0.14	0.32
Land Cultivated Status								
Marginal	0.15	0.58	0.23	0.57	0.33	0.18	0.24	0.32
Small	0.15	0.57	0.28	0.61	0.34	0.27	0.27	0.36
Medium	0.17	0.57	0.26	0.62	0.33	0.27	0.29	0.36
Semi-medium	0.18	0.56	0.28	0.63	0.36	0.26	0.25	0.36
Large	0.17	0.56	0.26	0.67	0.35	0.23	0.19	0.35
Poverty Status								
Non-poor	0.16	0.56	0.25	0.66	0.34	0.25	0.25	0.35
Poor	0.18	0.57	0.28	0.58	0.36	0.27	0.27	0.36
Tenancy status								
Owner	0.17	0.56	0.26	0.63	0.35	0.25	0.26	0.35
Tenant	0.15	0.61	0.49	0.63	0.37	0.31	0.24	0.40
State								
Telangana	0.20	0.57	0.17	0.61	0.34	0.23	0.28	0.34
Karnataka	0.12	0.54	0.28	0.70	0.32	0.24	0.23	0.35
Andhra Pradesh	0.19	0.58	0.35	0.57	0.38	0.30	0.25	0.37
Total	0.17	0.56	0.27	0.63	0.35	0.26	0.26	0.35

When we looked into the data further, we found a significant variation in the FDI at the district level that made us think and prioritize it at the sub-district level to capture variations at the smaller level (Said et al., 2011; Murthy et al., 2014; Patnaik et al., 2013; Kaul and Thornton, 2014). So, we focused on developing the FDI at the sub-district level, specifically at the mandal/block level (Table 16.3). This would not only ensure addressing the specific challenges faced by each area but also help find areas with high FDI scores and understand the reasons behind them (Khan and Salman, 2012; Said et al., 2011). After having the final result, we could feel the severity of the situation. So, we categorized the mandals/blocks into three groups based on their FDI values.

TABLE 16.3
Sub-District Level (Block Wise) Farmers Distress Index (FDI)

State/Block	Exposure	Adaptive Capacity	Sensitivity	Adaptation	Trigger	Psychology	Impact	FDI
Telangana								
Dornakal	0.24	0.58	0.14	0.60	0.33	0.29	0.32	0.36
Duggondi	0.12	0.56	0.22	0.48	0.39	0.13	0.49	0.34
Mhahboobabad	0.25	0.55	0.21	0.67	0.34	0.29	0.28	0.37
Nallabelli	0.16	0.57	0.13	0.60	0.40	0.28	0.21	0.33
Narsampet	0.17	0.56	0.11	0.58	0.35	0.23	0.31	0.33
Tirumalayapalem	0.20	0.57	0.24	0.60	0.33	0.22	0.25	0.34
Wyra	0.20	0.59	0.15	0.69	0.29	0.17	0.20	0.33
Total	**0.20**	**0.57**	**0.17**	**0.61**	**0.34**	**0.23**	**0.28**	**0.34**
Karnataka								
Ballari	0.17	0.54	0.21	0.63	0.38	0.29	0.28	0.36
Maski	0.12	0.53	0.41	0.70	0.25	0.26	0.28	0.36
Shahapur	0.11	0.54	0.26	0.72	0.30	0.22	0.15	0.33
Shorapur	0.12	0.54	0.30	0.72	0.30	0.19	0.20	0.34
Siruguppa	0.13	0.56	0.23	0.68	0.34	0.25	0.22	0.35
Sirwar	0.09	0.54	0.29	0.72	0.33	0.22	0.26	0.35
Total	**0.12**	**0.54**	**0.28**	**0.70**	**0.32**	**0.24**	**0.23**	**0.35**
Andhra Pradesh								
Amaravathi	0.21	0.59	0.39	0.60	0.39	0.28	0.19	0.38
Darsi	0.21	0.58	0.20	0.53	0.35	0.23	0.27	0.34
Durgi	0.21	0.57	0.34	0.56	0.36	0.32	0.26	0.37
Kanchikacherla	0.14	0.62	0.35	0.57	0.40	0.26	0.19	0.36
Kankipadu	0.08	0.60	0.25	0.65	0.41	0.35	0.28	0.37
Kurichedu	0.21	0.56	0.42	0.55	0.43	0.29	0.27	0.39
Veerullapadu	0.17	0.59	0.40	0.58	0.38	0.38	0.26	0.40
Veldurthi	0.13	0.56	0.37	0.56	0.42	0.42	0.25	0.39
Total	**0.23**	**0.56**	**0.40**	**0.55**	**0.35**	**0.21**	**0.38**	**0.38**

The first category, marked with a red symbol, represented severe mandal/block distress. These were the top 30% of the mandals/blocks with the highest FDI scores. The second category, marked with a yellow symbol, represented moderate mandal/block distress, consisting of 30–60% of the mandals/blocks. And, finally, the third category, marked with a green symbol, represented low mandal/block distress, consisting of the bottom 40% of the mandals.

Our research not only highlighted the challenges faced by the farmers but also emphasized the need for urgent action to provide them with the support they desperately needed. Though we should remember that the FDI was just one aspect of a multi-dimensional approach, we were hopeful that our research could make a positive change and improve the lives of these poor farmers.

REFERENCES

Kaul V, Thornton TF. (2014). Resilience and Adaptation to Extremes in a Changing Himalayan Environment. *Regional Environmental Change*, 14, pp. 683–698.

Khan FA, Salman A. (2012). A Simple Human Vulnerability Index to Climate Change Hazards for Pakistan. *International Journal of Disaster Risk Science*, 3(3), pp. 163–176.

Murthy CS, Laxman B, Sesha Sai MVR, Diwakar PG. (2014). Analysing Agricultural Drought Vulnerability at Sub-District Level through Exposure, Sensitivity and Adaptive Capacity Based Composite Index. *The International Archives of the Photogrammetry, Remote Sensing and Spatial Information Sciences*, 40, p. 65–70.

Patnaik U, Das PK, Bahinipati CS. (2013). Analyzing Vulnerability to Climatic Variability and Extremes in the Coastal Districts of Odisha, India. *Review of Development and Change*, 18(2), pp. 173–189.

Reddy AA, Bhattacharya A, Reddy SV, Ricart S. (2021). Farmers' Distress Index: An Approach for an Action Plan to Reduce Vulnerability in the Drylands of India. *Land*, 10(11), p. 1236.

Said F, Musaddiq T, Mahmud M. (2011). Macro Level Determinants of Poverty: Investigation through Poverty Mapping of Districts of Pakistan. *The Pakistan Development Review*, 50(4), pp. 895–910.

17 Scaling Up of Antifragile Strategies

OUR FINAL THOUGHTS

Finally, it was the wrap-up time of the research. After spending months studying the issue of agriculture distress, talking to countless farmers, observing their daily struggles, and analysing data, we understood the root causes of the farmers' crisis and its devastating impact on farmers and rural communities. Through this research, we have come to realize that it is a complex issue that is difficult be solved with a uniform approach. It is a mix of factors that are all tangled up and making each other worse.

But amidst all this, we found something hopeful. Farmers have been using traditional ways to cope with these challenges for ages, which have helped them maintain a stable income and ensure food security even during times of distress. But with the changing times and increasing commercialization of agriculture, these practices alone are not enough. It is crucial to integrate traditional practices with modern technologies and innovations to make agriculture more sustainable and resilient to risks. And while the government has tried to help with things like crop insurance, it has not always worked out perfectly. So, to really fix things, we need to look at the big picture. That means giving farmers better access to irrigation facilities, promoting scientific farming practices, providing timely credit and marketing support, and ensuring fair prices for agricultural produce. By taking these steps, we can ensure a prosperous and resilient future for farming communities and ecosystems.

MAIN DISCOVERIES OF OUR RESEARCH

We started our study with great determination and a sense of duty in a specific area. We knew that having updated information was crucial to tackling climate-related risks and reducing distress among farmers (Kantamaneni et al., 2020; Ghosh, 2018). Our goal was to create a Farmers Distress Index (FDI) to help policymakers address farmers' problems effectively (Gulati et al., 2018) and evolve measures to tackle them (Simane, 2011). After months of hard work, we identified 50 key factors across seven categories that were important for understanding farmers' vulnerability. These included things like how much risk they faced, how well they could adapt, and how their strategies worked. The final results of our study were eye-opening. FDI scores varied a lot across different locations, showing that some areas were much more distressed than others. This made it clear that a targeted approach was needed to help farmers in each place. We also found that, within each district, sub-district areas had different FDI scores. This meant that we needed to focus on a smaller level to really help farmers. As we dug deeper, we realized that the main cause of distress among

DOI: 10.1201/9781003582878-17

farmers was the climate crisis. Unpredictable weather and natural disasters were hitting farmers hard, affecting not just their money but also their social lives. This made us feel an urgent need to share our findings and work towards helping farmers. Our study gave us important insights and showed us that action needed to be taken to support our farmers and keep our agriculture sustainable.

A JOURNEY OF TRIALS

However, our research project threw some of the toughest challenges that we have ever faced. First, the COVID-19 pandemic hit, making it hard to get consistent government data because of lockdowns. Conducting field surveys in villages took a lot of time, and we had to make sure everyone stayed safe. Also, we could not cover as many areas as we wanted due to limited resources and time. But we were determined that we would not let any obstacles stand in our way. We kept pushing through because we knew how important our topic was, not just for farmers, but for society as a whole. With climate change and changes in farming, finding a solution was more important than ever.

WHAT'S NEXT? OUR DIRECTION FOR THE WAY FORWARD

Finally at the end of the research, with the FDI in hand, our team has some exciting recommendations for future work (Reddy et al., 2021). Our main idea is to create better agricultural insurance schemes. We realize that instead of just focusing on one crop, these new schemes can cover a whole range of risks, giving farmers more protection. This would not only provide financial stability for the farmers but also encourage them to take risks and experiment with new crops.

But that is not all. We also thought about how the FDI could help target the areas that need the most help. By identifying the most distressed areas, we can make sure that support and policies reach the farmers. This can make a significant difference in improving their lives. One of our most brilliant ideas is to set up "Climate Innovation Platforms" in vulnerable districts. These would be like think tanks where local communities can come together to plan for the future and come up with ways to tackle climate challenges. This way, we can involve the farmers in finding solutions and empower them to take control of their future.

We are confident that our recommendations can make a real impact, especially in places like Maharashtra, Telangana, Karnataka, and Rajasthan where farmer distress is a big issue. But we are also aware that our research was just the beginning. To make these ideas a reality, we require a complete distress management package, which means we need to imagine big and have a clear action plan, a budget, and a lot of teamwork between researchers, policymakers, and organizations (Timilsina et al., 2020; Cariappa et al., 2021; Cattivelli and Rusciano, 2020; Rusciano et al., 2017, 2020) (Table 17.1; Figure 17.1).

And this is where everybody's support is required. Together, we can create a brighter future for farmers and build more sustainable and resilient communities. Our work may have started with just a tool, but with determination and collaboration, we can make a lasting difference in the lives of farmers and their families.

TABLE 17.1
Action Plan of Distress Management Package (an Example)

Examples of Distress Indicators	Area of Intervention	Action Plan for Intervention	Responsible Bodies
Lack of non-farm employment	Strengthening and training of small enterprises	Encouragement of women/youth in engaging in cottage industry with farm waste materials	NGOs and skill development council
High indebtedness	Credit support	Easy and smooth access of formal credit institution	Formal credit institutions (banks, cooperatives)
Low agricultural land landholding	Mapping of local resources and their management, development of land-lease markets	Adoption of integrated farming system, credit facilities to tenant farming	Local bodies Banks
Low % irrigated area (lack of irrigation)	Watershed development	Practice of water harvesting and conservation	Local bodies Community
Low educational status	Capacity building	Provision of extension services and special training to farmers	Extension agencies Research institutions NGOs, SHG
Crop failure	Promotion of involvement in the mitigation programme	Awareness of crop insurance schemes Identification of drought-prone areas Selection of proper crop varieties	Local-, state-, and national-level government bodies Insurance companies

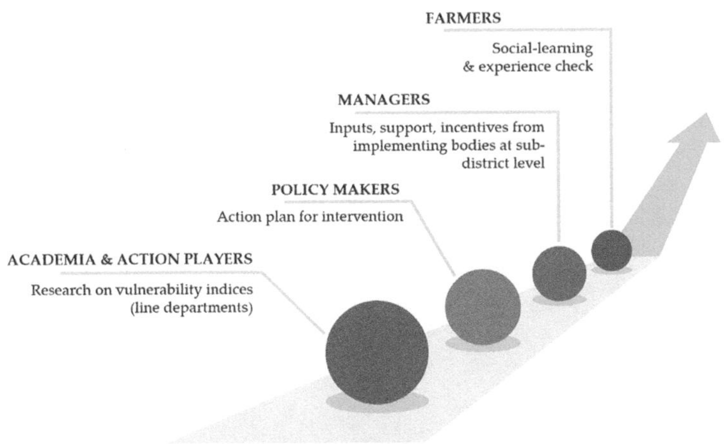

FIGURE 17.1 Diagrammatic representation of distress management package in agriculture.

REFERENCES

Cariappa AA, Acharya KK, Adhav CA, Sendhil R, Ramasundarma P. (2021). Impact of COVID-19 on the Indian Agricultural System: A 10-Point Strategy for Post-Pandemic Recovery. *Outlook on Agriculture*, 50 (1), pp. 26–33.

Cattivelli V, Rusciano V. (2020). Social Innovation and Food Provisioning during COVID-19: The Case of Urban–Rural Initiatives in the Province of Naples. *Sustainability*, 12 (11), pp. 4444.

Ghosh RK. (2018). *Performance Evaluation of Pradhan Mantri Fasal Bima Yojana (PMFBY). Part 1. Governance Analysis*. Final Report. Indian Institute of Management Ahmedabad, Ahmedabad, India, pp. 85–86.

Gulati A, Terway P, Hussain S. (2018). *Crop Insurance in India: Key Issues and Way Forward*. Working Paper No. 352. Indian Council for Research on International Economic Relations, New Delhi, India,

Kantamaneni K, Rice L, Yenneti K, Campos LC. (2020). Assessing the Vulnerability of Agriculture Systems to Climate Change in Coastal Areas: A Novel Index. *Sustainability*, 12 (11), p. 4771.

Reddy AA, Bhattacharya A, Reddy SV, Ricart S. (2021). Farmers' Distress Index: An Approach for an Action Plan to Reduce Vulnerability in the Drylands of India. *Land*, 10(11), p. 1236.

Rusciano V, Civero G, Scarpato D. (2017). Urban Gardens and Environmental Sustainability: An Empirical Research of Campania Region. Quality-Access to Success, 18, pp. 376–381.

Rusciano V, Civero G, Scarpato D. (2020). Social and Ecological High Influential Factors in Community Gardens Innovation: An Empirical Survey in Italy. *Sustainability*, 12 (11), p. 4651.

Simane B. (2011). Building Resilience to Climate Change and Green Economy in Mountain Ecosystems of Ethiopia: Integrating Research, Capacity Building and Sustainable Development Activities. *Proceedings of the Stakeholders Workshop*. Debre Markos, Ethiopia, p. 92.

Timilsina B, Adhikari N, Kafle S, Paudel S, Poudel S, Gautam D. (2020). Addressing the Impact of COVID-19 Post-Pandemic on Farming and Agricultural Deeds. *Asian Journal of Advanced Research and Report*, 11 (4), pp. 28–35.

Index

Note: **Bold** page numbers refer to tables and *italic* page numbers refer to figures.

For Product Safety Concerns and Information please contact our EU representative GPSR@taylorandfrancis.com Taylor & Francis Verlag GmbH, Kaufingerstraße 24, 80331 München, Germany